Production and
Management of
Cultivated Forages

Production and Management of Cultivated Forages

PETER D. WALTON

Professor, formerly Chairman
Department of Plant Science
University of Alberta
Edmonton, Alberta

RESTON PUBLISHING COMPANY, INC., RESTON, VIRGINIA
A Prentice-Hall Company

Library of Congress Cataloging in Publication Data

Walton, Peter D.
  The production and management of cultivated forages.

  Includes index.
  Bibliography: p.
  1. Forage plants. I. Title.
  SB193.W26 1982   633.2   82-12306
  ISBN 0-8359-5622-9

Editorial/production supervision and interior design
by Barbara J. Gardetto

© 1983 by
Reston Publishing Company, Inc.
A Prentice-Hall Company
Reston, Virginia 22090

All rights reserved. No part of this book may be reproduced in any way, or by any means, without permission in writing from the publisher.

10 9 8 7 6 5 4 3 2 1

Printed in the United States of America

This book is dedicated to
 Joyce,
who shared these many years with me

Table of Contents

Preface xv

chapter 1
THE NATURE OF FORAGE PLANTS 1
    Introduction 1
    Grasses 3
        Shoot System 4
        Root System 5
        Inflorescence 7
        The "Seed" 11
    Legumes 12
        Stems and Leaves 12
        Roots 12
        Inflorescence 12
    Further Reading 15

chapter 2
HISTORIC PERSPECTIVE 16
    Introduction 16
    Geology 16
    Beginning of Farming 18

Farming in Europe 19
Farming in North America 20
Further Reading 23

chapter 3
THE GRASSES 24

Introduction 24
Cool-Season Perennial Grasses 25
    Bentgrasses (genus *Agrostis*) 25
    Bluegrasses (genus *Poa*) 25
    Smooth Bromegrass (*Bromus inermis* L.) 29
    Fescues (genus *Festuca*) 32
    Meadow Foxtail (*Alopecurus pratensis* L.) 36
    Orchardgrass (*Dactylis glomerata* L.) 39
    Reed Canarygrass (*Phalaris arundinaceae* L.) 41
    Ryegrass (genus *Lolium*) 43
    Wildrye (genus *Elymus*) 45
    Timothy (*Phleum pratense* L.) 50
    Wheatgrasses (genus *Agropyron*) 52
Warm-Season Perennial Grasses 65
    Bahiagrass (*Paspalum notatum* Fliigge) 67
    Bermudagrass (*Cynodon dactylon* L. Pers.) 69
    Carpetgrass (*Axonopus affinis* Poir) 72
    Dallisgrass (*Paspalum dilatatum* Poir) 72
    Johnsongrass (*Sorghum halepense* L. Pers.) 74
    Pangolagrass or Digitgrass (*Digitaria decumbens* Stent.)
    Rhodesgrass (*Chloris gayana* Kunth) 77
    Saint Augustinegrass (*Stenotaphrum secundatum* Walt.) 77
Further Reading 77

chapter 4
THE PERENNIAL LEGUMES 79

Species Used in Temperate Areas 79
    Alfalfa (*Medicago* spp.) 79
    Birdsfoot Trefoil *(Lotus corniculatus)* 83
    True Clovers (genus *Trifolium*) 84
    Cicer Milkvetch (*Astragalus cicer* L.) 93
    Sainfoin (*Onobrychis* spp.) 95
    Sweetclover (*Melilotus* spp.) 97
    Vetches (*Vicia* spp.) 99
Warm-Climate Legumes 99
    Arrowleaf Clover *(Trifolium vesiculosum)* 100
    Crimson Clover *(Trifolium nicaenatum)* 101
    Kudzu (*Pueraria lobata* Willd.) 101

Annual and Perennial *Lespedeza* Species   102
Subterranean Clover *(Trifolium subterraneum)*   102
Further Reading   106

chapter 5
ANNUAL FORAGES   108
Introduction   108
Why Annuals?   110
Uses   111
Mixtures   114
Crop Residues   114
Crop Species Used as Annual Forages   115
    Large-Grain Crops   116
    Small-Grain Crops   117
    Annual Grasses   117
    Legumes   119
    Succulent Fodders   119
Further Reading   120

chapter 6
ESTABLISHMENT   121
Introduction   121
Choice of Crops   123
Selection of Mixtures   123
Making a Forage Mixture   125
Seed Quality   125
Germination   126
Seedling Vigor   128
Seedbed and Seeding Depth   130
Seeding Equipment   132
Time of Planting   132
Seed Rates   133
Seeding Pattern   133
Seed Treatment   134
Companion Crops   139
Grassland Renovation   140
Further Reading   143

chapter 7
THE PHYSIOLOGY OF FORAGE CROP GROWTH: THE GRASSES   144
Plant Foods   144
Carbohydrate Reserves   146

Storage Regions 148
Relationships Between Storage Substances and Species 148
Management of Carbohydrate Reserves 149
    Temperature and Carbohydrate Reserves 149
    Water and Carbohydrate Reserves 149
    Nitrogen and Carbohydrate Reserves 150
Growth and Regrowth 150
Defoliation of a Tiller 154
    Seasonal Changes in Growth Rates 155
Winter Dormancy 157
Summer Dormancy 158
Plant Stress 158
    Low Temperatures 158
    High Temperatures 159
    Water Deficits (Drought) 159
    High Salt Concentrations 159
Further Reading 160

chapter 8
THE PHYSIOLOGY OF FORAGE CROP GROWTH: THE LEGUMES 161

Introduction 161
Shoot Growth 162
Root System 164
Growth Responses 164
Further Reading 166

chapter 9
FORAGE QUALITY 168

Introduction 168
Forage Chemistry 169
Nitrogenous Compounds 169
Nonprotein Nitrogenous Compounds 170
Carbohydrates, Including Pectin 170
    Lignin 171
    Lipids 171
    Organic Acids 171
    Pigments 172
    Vitamins 172
    Minerals 172
Proximate Feedstuff Analysis 173
    Crude Protein (CP) 173
    Crude Fiber (CF) 174

Acid-Detergent Fiber 175
Review 175
Energy 175
Digestibility 176
Intake 176
Variation in Forage Quality 178
Climate and Quality 182
Fertilization and Quality 182
Influence of Management on Quality 183
Further Reading 183

chapter 10
ANTIQUALITY FACTORS 185

Introduction 185
Cyanogenetic Glucosides 186
Saponins 187
Bloat 187
Sweetclover Bleeding Disease of Cattle 189
Tannins 190
Flavonoids 190
Alkaloids 191
Tall Fescue Toxicity 192
Nitrate Poisoning 193
Grass Tetany 194
Further Reading 194

chapter 11
FORAGE STORAGE: DRY SYSTEMS 195

Introduction 195
Hay-Making Principles 196
Hay-Making Methods 199
    Long Hay 199
    Chopped Hay 202
    Shredded Hay 203
    Baled Hay 203
    Large Bales 203
    Barn Drying 208
    Loose Hay Systems 209
    Wafered and Pelleted Hay 209
    Hay Additives 210
Summary 211
Dehydration 212
Further Reading 213

chapter 12
FORAGE STORAGE: SILAGE 215

Introduction 215
Process of Bacterial Fermentation 216
Silage-Making Problems 217
    Too Much Air 218
    Too Much Water 218
    Too Little Carbohydrate 219
Methods of Making Silage 219
    Direct-Cut Silage 219
    Wilted Silage 221
    Low-Moisture Silage (Haylage) 221
Chemical Additives 222
Harvest Equipment 223
Types of Silos 225
Summary 226
Further Reading 228

chapter 13
THE USE OF FERTILIZERS 229

Past Use and Present Needs 229
Nutrient Requirements 230
    Grasses 230
    Legumes 235
    Grass-Legume Mixtures 237
    Plant and Animal Waste 238
Nitrogen Balance in Pastures 240
Further Reading 243

chapter 14
THE PESTS OF FORAGE CROPS 244

Losses Due to Weeds 244
Characteristics of Weeds 245
Poisonous Weeds 247
Weed-Control Methods 248
    Direct Destruction 249
    Ecological Control 249
    Biological Control 249
    Chemical Control 250
    Nonselective Herbicides 250
    Selective Herbicides 250
Avoiding Infestations 252
Insect Pests 252
    Types of Insects Attacking Forage 252

Insect Control 254
    Cultural Control 254
    Resistant Cultivars 255
    Biological Control 255
    Chemical Control 255
Forage Crop Diseases 256
Diseases of Legumes 256
    Bacterial Diseases 256
    Fungus Diseases 256
    Virus and Mycoplasma Diseases 257
Legume Diseases in Southeastern United States 258
Diseases of Grasses 258
    Warm-Season Grasses 258
    Cool-Season Grasses 259
Disease Control 260
Further Reading 261

chapter 15
SEED PRODUCTION FROM LEGUMES AND GRASSES 262

Introduction 262
Pollination Mechanisms 263
    Grasses 263
    Legumes 265
    Leaf-Cutter Bees 269
Seed Trade Organizations 270
    Plant-Breeding Practices 270
    Seed Multiplication 273
    Commercial Seed Production Practices 274
Further Reading 275

chapter 16
PALATABILITY AND GRAZING BEHAVIOR 276

The Animal's Use of Its Senses 276
    Smell 276
    Touch 277
    Taste 277
Evidence from Experiments and Experience 277
    The Individual Animal and the Herd 278
    Plant Species and Their Growth Stage 279
    Choice of Plant Parts 279
    Influence of Previous Diet 279
    Management 280
    Quality Characteristics of Forage 280
Grazing Behavior 280

Herd Behavior 281
Further Reading 282

chapter 17
PASTURE MANAGEMENT 283

Economic Importance of Pastures 283
Art of Management 284
Forage Yield Variation 284
Stocking Rate 288
Plant Growth 289
Plant Characters 289
Undergrazing and Overgrazing 289
Review 290
Grazing Systems 290
Rotational Grazing 292
    Example of Rotational Grazing 293
    Changes in Botanical Composition 296
    Essential Features of Rotational Grazing 299
Variations in Rotational Grazing Methods 299
Zero Grazing 300
Stockpiling 302
Conclusion 302
Further Reading 303

Appendices 305

Index 314

Preface

The energy crisis, and the escalating feed grain prices which accompany it, has now placed an ever-increasing emphasis on the importance and use of forage material in all its many forms. This contrasts with the situation in the past. In all countries of the world, research and development in forage production, management, and utilization have been neglected relative to the effort and attention given to other crops. This is indeed unfortunate, since forage production is a truly complex multidisciplinary topic which must draw on the basic findings of many scientific disciplines for successful implementation. From remote antiquity, before the known presence of people in the world, herbivores have utilized our present grassland areas. Such zones now represent an ancient renewable resource of proven worth to humanity. Our grasslands are capable of producing indirectly proteins and fats in more abundance, and more economically, than any other sector of the agricultural industry. In a world with a rapidly expanding population, such a valuable resource calls for careful study, appreciation, and understanding.

The object of this book is to integrate information from many different scientific fields into a composite management and production package which will be of value to all those who seek to make "two blades of grass grow . . . where only one grew before" [Jonathan Swift (1667–1765); from *Gulliver's Travels*, Voyages to Brobdingnag, ch. 7]. The book presents, concisely, fundamental information concerning forage plant species and their cultivars; it considers the various ways in which these plants are conserved and the chemical and physical characteristics which constitute forage quality. It sets out the manner in which quality is determined and discusses the detrimental characteristics of forage crops. The fundamentals of whole-plant physiology as they apply to forage crops are presented and related to the management practices involved in grazing, as well as hay and silage making. The establishment of perennial and annual forage plants for pasture, conservation for winter feed, or seed production is studied. Reasons for the seed trade being organized as it is and the relationship between it and both the plant breeder and the farmer are considered. The use of fertilizers and the methods and importance of weed control also receive attention.

The material presented in the book is based on the contents of two half-classes offered by the Department of Plant Science at the University of Alberta for which the author is the instructor. The material included was collected and assembled during some 14 years of study, field and library research, and practical experience. Over these many years, the author has drawn freely from extensive and varied sources. Scientific papers, bulletins from provincial and federal departments of agriculture in both Canada and the United States, and many books have all been used. Valuable discussions have led to much useful advice being obtained from students, faculty, and staff at the University of Alberta, as well as from farmers, ranchers, representatives of the forage industry and the seed trade, and from the staff of federal and provincial departments of agriculture in Canada, in the United States, and in many other nations.

The author wishes to acknowledge this extensive and valuable assistance in assembling the material for this book. The author also wishes to thank Ms. Colleen Blais and Mr. Hans Jahn for reading the text and for their many helpful comments and suggestions. The illustrations and line drawings of the forage plants were done by Mr. John Reading Maywood. They provide an accurate visual representation more valuable than a lengthy verbal account. Some of the material set out in Chapter 2 was previously published by the author in *Economic Botany*. The author wishes to thank the curator of publications, New York Botanical Gardens, for permission to use it here. Some of the information in Chapter 3 concerning smooth bromegrass was previously published by the author in Volume 33 of *Advances in Agronomy* and he wishes to thank the Academic Press, Inc., for permission to present it in this book. Finally, a special word of thanks is due to Ms. Mary Douglas Wright for the care and attention to detail which she has given in reading and in several rereadings of the text.

Peter D. Walton
April 1981

chapter 1

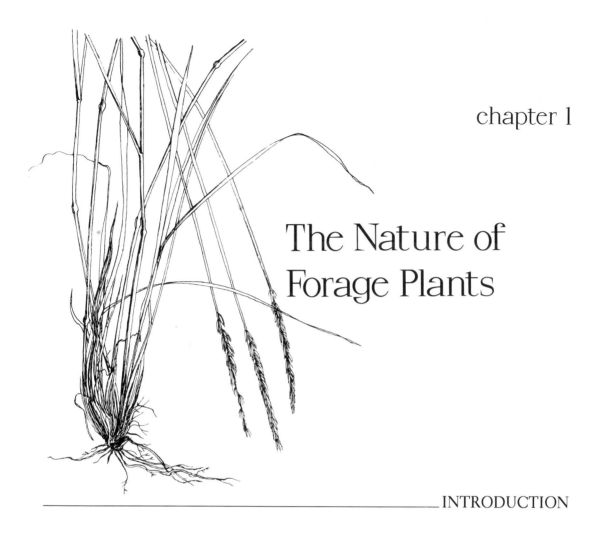

The Nature of Forage Plants

INTRODUCTION

The study of forage crops is one of the most fascinating aspects of plant science. It involves an appreciation of the work of specialists in many scientific fields; mathematicians, physicists, chemists, economists, engineers, entomologists, physiologists, pathologists, biochemists, and zoologists, as well as botanists, all make important contributions. The production and management of cultivated forages is, in fact, a truly multidisciplinary topic, which is filled with both challenges and rewards. The diversity we find in our study of forages is the outcome of the interaction between the soil, the plant, and the animal, and it includes all the factors associated with the growing, grazing, conserving, storing, and feeding of forage substances.

Grasslands, which produce the major part of the world's forage, comprise about 25% of the world's vegetation area (see Chapter 2). They occupy by far the largest area of any single plant type. Such areas form well over 50% of the total land mass in many countries. It is, however, difficult to assess their economic importance or their productivity, because accurate yield *data* from which national output may be determined are frequently lacking. In the United States, for example, neither the agricultural census nor the federal-state crop-reporting service provides information on the productivity of grasslands, except as harvested for hay or silage. The same situation exists in Canada. Even where hay and silage production is recorded in national or provincial surveys, the extent of direct "on farm" usage is questionable and may, or may not, be included. Furthermore, it is difficult to determine, from the statistics presented by government agencies, what information applies to native as opposed to cultivated forage plants.

The prime value of our grassland areas lies in the support they provide for our livestock industry. Nonruminant animals, such as horses, pigs, and poultry, are able to use vegetative plant material as a part of their diets. Ruminant animals, however, may be supported entirely by fibrous forage material which is digested by the large and varied population of microorganisms in their rumens. Forages have been defined as *plant material with a dry matter fiber content over 25%*. Percentage of fiber values much higher than this are frequently encountered, as we will see later in this text in Chapter 9, where forage quality and feed value will be discussed in detail.

While the prime importance of forage lies in the economic value of the meat, milk, or, in some cases, work which is produced from it, forage plants also make an important contribution to soil fertility. Forages can maintain or improve soil fertility and structure, as well as protect it from both wind and water erosion. Where row-crop farming has led to soil deterioration, forage crops are commonly used to restore and stabilize the land. In so doing, the grass cover is of value in the conservation of water for both agricultural and domestic use. Forage grasses also have an esthetic appeal and are frequently used to cover, protect, and beautify yards, playgrounds, parks, highways, and airfields, and to provide feed for wildlife.

In spite of these many and important uses, forage plants have received far less attention from farmers and research workers than have row crops. Studies using annual row crops have drawn attention to the importance of the appropriate use of fertilizers and of high plant populations. This, together with improved cultivars, has made row cropping very remunerative. As a result, agricultural development has been such that annual crop production has replaced natural prairie grassland, while perennial grasslands now occupy areas which formerly were forest. These latter areas are often hilly and of low productivity. They are used for forage production because either the machines used to establish and harvest row crops do not function on such topography, or because soil fertility is low and high annual crop yields are not expected.

These circumstances have led to increased grain production and, hence, low grain prices, so farmers have used grain rather than forage to feed their livestock. Drastic increases in energy costs have reversed this situation. Grain production is much more energy dependent than forage production, so when fuel costs rise, grain prices increase more rapidly than those for forage. These high costs, combined with an ever-expanding world population, have led many individuals to question the propriety of feeding grain to animals when part of our human population does not have enough to eat. Thus, a combination of economic and social pressures has resulted in an increasing use of forage material.

Bulky fibrous plant material is, then, of ever-increasing importance for livestock production. What kinds of plants provide such material? They come from two large botanical families, the grasses and the legumes.

GRASSES

The grasses, which are herbaceous, monocotyledonous plants with jointed stems and sheathed leaves, are the outcome of several million years of evolutionary adaptation to growth and reproduction in the presence of the grazing animal. In the vegetative stage, stem and buds are close to the ground, the aerial parts of the plant consisting almost entirely of leaves and leaf sheaths (Figure 1-1).

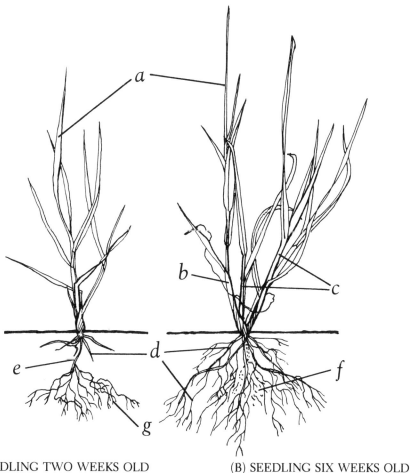

(A) SEEDLING TWO WEEKS OLD   (B) SEEDLING SIX WEEKS OLD

- a. leaves (lamina)
- b. tiller
- c. leaf sheaths
- d. adventitious roots
- e. epicotyl
- f. degenerating seminal roots
- g. seminal roots

FIGURE 1-1
Nonelongated grass plants.

Shoot System

The grass plant is composed of a number of shoots, or *tillers*, which in turn are composed of *leaves* and *leaf sheaths*. The leaf sheaths are enclosed one inside the other, so that they appear to form a "stem." The oldest leaf is the one which is lowest on the plant, while new leaves emerge from the top and center of the cylindrical column of leaf sheaths. At the junction of the leaf blade and the leaf sheath are frequently present two structures, the *ligule* (Figure 1-2) and the *auricle* (Figure 1-3), which vary widely in form between grass species. These structures are a valuable means of species identification and are widely used in grass classifications.

Grass shoots arise in two ways: from the embryo in the seed or from a vegetative bud in the leaf axis. When the plant is in the vegetative stage, the true stem has very short internodes and is held well below the bite level of the grazing animal. This is important, since the tip of the stem consists of an intensely active meristematic area from which all growth originates. Early in the growing season, this *apical dome* produces leaf buds which, when they develop, have in their axil another bud which is capable of developing into a new tiller. When environmental conditions are favorable,

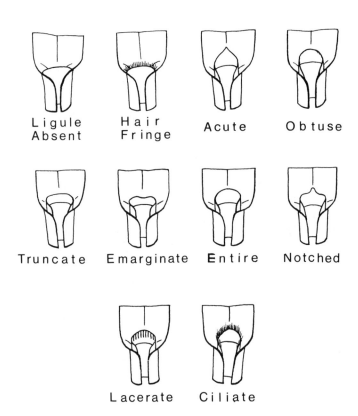

FIGURE 1-2
Types of ligules found in grasses.

Auricles Absent   Clawlike   Rounded   Rudimentary

FIGURE 1-3
Types of auricles found in grasses.

the plant is able to produce new tillers which are not only identical to those from which they arise, but are themselves capable of *tillering*. Thus, the grass plant might spread vegetatively and indefinitely. There are, in fact, reports of individual creeping red fescue plants which cover an area over 1 kilometer in diameter.

The grass plant will respond to appropriate temperature and light conditions (see Chapter 15) by producing, from the apical dome, buds which will develop into flowering heads, rather than vegetative structures. At this time, the internodes elongate very rapidly. The leaves which have been held together in a tight rosette (Figure 1-1) are separated vertically, and the flowering head is pushed up and out of the cylindrical column formed by the leaf sheaths (Figure 1-4). The grass stem produced in this way is called a *culm*. It is at this stage that the production of dry matter, or plant bulk, is most rapid. To take advantage of this rapid and substantial increase in mass, hay crops are normally cut after the head has emerged and at about the time of anthesis.

Branching at the base of the culm produces new tillers which behave in one of two ways. The new tillers will either grow up within the leaf sheath, in which case the species will have a tufted or *bunch* appearance [Figure 1-5(A)] or they will break through the leaf sheath to produce new horizontal stems and a plant with a "creeping" habit [Figure 1-5(B)]. These stems may remain above ground [*stolons*, Figure 1-6(C)] or be below ground level [*rhizomes*, Figure 1-6(B)]. Both types of stem are able to produce new tillers from vegetative buds at their nodes.

Root System

Grasses have two fibrous root systems, each consisting of numerous thin roots. The *seminal* roots develop from primordia which are present in the embryo. This root system is highly branched but very short lived. It exploits a large soil volume during the early stage of the plant's development (see g in Figure 1-1). The second or *adventitious* root system is developed from the lower nodes of each tiller. In the perennial grasses, this is the only root system available after establishment. In the case of annual grasses, the seminal root system may function throughout the life of the plant.

As well as performing the normal root function of nutrient absorption, the numerous adventitious roots of the perennial grasses are important in anchoring the plant in the soil. They enable the plant to withstand the tug and pull of the grazing animal. The presence of these numerous and extensive adventitious roots is also important in the accumulation of organic matter under a grass sward. Each tiller develops its own root system. When the tiller dies (i.e., whenever the apical dome is removed by cutting or grazing, or when flowering takes place), the root system may die with it. As a result, organic matter accumulates near the soil surface, since the root systems in many grass species are relatively shallow.

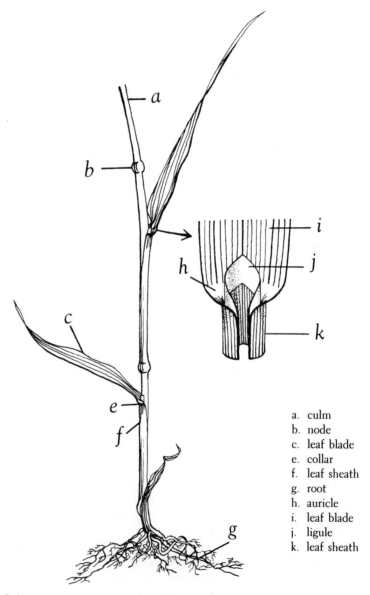

FIGURE 1-4
Elongated grass plant.

a. culm
b. node
c. leaf blade
e. collar
f. leaf sheath
g. root
h. auricle
i. leaf blade
j. ligule
k. leaf sheath

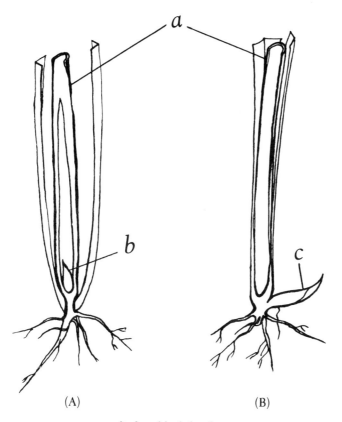

a. cylinder of leaf sheaths
b. bud growing up inside leaf sheath
c. bud growing through leaf sheath

FIGURE 1-5
(A) Tiller growing up within leaf sheath in a tufted species and (B) breaking through leaf sheath to form a stolon or rhizome.

Inflorescence

The simplest type of infloresence is a *raceme* in which the *spikelets* are borne on an unbranched axis. There are two types of grass inflorescence which occur most commonly. One is the *spike* [Figure 1-7(A)], which has sessile spikelets on an unbranched *rachis* or main axis. The other is the *panicle* [Figure 1-7(B)], in which the rachis has at least a second order of branching, frequently extensive. Between the two extremes, there exists a great range of diverse and intermediate types. All inflorescences are, however, built up from the same basic structures in exactly the same way.

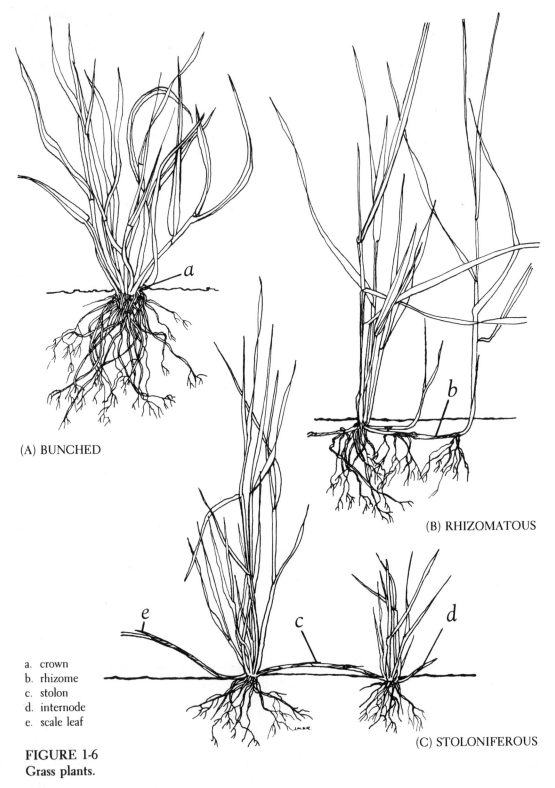

a. crown
b. rhizome
c. stolon
d. internode
e. scale leaf

FIGURE 1-6
Grass plants.

The smallest of the basic structures making up the inflorescence is the *floret* [Figure 1-8(E) and (D)], which is enclosed by the *lemma* and the *palea*. Florets are grouped together to form *spikelets,* enclosed by the *glumes* [Figure 1-8(B) and (C)]. The spikelets, in turn, are the structures which form the inflorescence [Figure 1-8(A)].

It is the great diversity of rachis length, as well as the number, size, and arrangement of the florets and spikelets, which results in the widely different appearances of grass inflorescences (see Chapter 3).

Within the floret, the floral parts nearly always consist of three *stamens,* two *styles* with feathery stigmas, and an *ovary* with a *single ovule* (Figure 1-8). Most grasses are cross-pollinated by the *wind.* Two small structures near the base of the lemma and palea (the *lodicules,* see *i* in Figure 1-8) swell up and open the florets, the anthers are extruded, and the stigmas exposed. There are, however, some self-pollinated grasses whose anthers burst before the floret opens.

(A) INTERMEDIATE WHEATGRASS—SPIKE

(B) ORCHARDGRASS— PANICLE

FIGURE 1-7
Types of grass inflorescence.

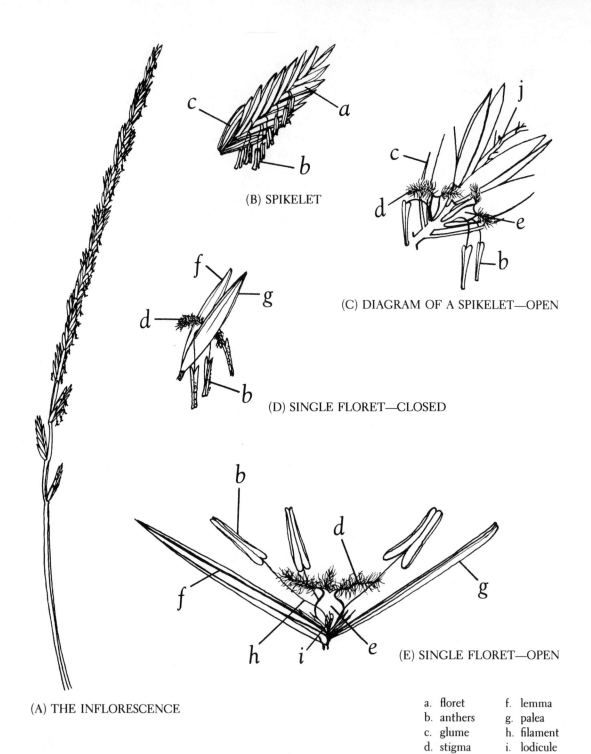

FIGURE 1-8
Components of the inflorescence.

The "Seed"

For the forage grasses, what is commercially called the "seed" is in fact the *caryopsis,* a type of fruit or ripened ovary, which is enclosed in the lemma and palea. The pericarp, or outer seed coat layer, is formed from the ovary wall. In a true seed, the ovary wall will form a pod (e.g., the legumes). In the grasses, the "seed" is formed when the fruit starts to swell and develop after fertilization has taken place. Simultaneously, the lodicules contract, closing the lemma and palea, which become fused to the wall of the developing caryopsis. The embryo is found in the caryopsis, next to the lemma [Figure 1-9(B)]. The caryopsis also contains the *endosperm,* a storage region for the food reserve substances on which the embryo will draw at the time of germination. In common with other plant species, the embryo consists of a *radicle* (root) and *plumule* (shoot) [Figure 1-9(A)].

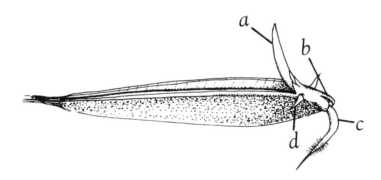

(A) GERMINATING GRASS SEED

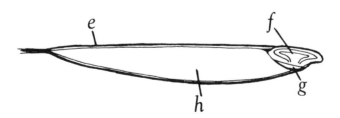

(B) SECTION THROUGH GRASS SEED

a. coleoptile
b. coleorhiza
c. first seminal root
d. second seminal root
e. aleurone layer
f. embryo
g. scutellum
h. endosperm

FIGURE 1-9

LEGUMES

The legumes differ markedly from the grasses. Botanically, they are dicotyledonous, have compound leaves (frequently with three or more leaflets), possess well-developed, and often deep, taproots, and are self- or insect-pollinated. They are widely grown, often in mixtures with grasses, which they complement in a number of ways. The most important of these, and the reason for the wide usage of legumes, is their ability to form a symbiotic relationship with bacteria of the genus *Rhizobium*. In association with the legume plants, the bacteria are capable of fixing atmospheric nitrogen. As a result of this association, legume plants supply the major part of the world's plant protein. The crude protein (see Chapter 9) content of legume leaves and stems is frequently found to be in the range of 12% to 24%. Legumes may be annuals, biennials, or, most frequently, perennials.

Stems and Leaves

Unlike the grasses, legumes have elongated internodes and a terminal growing point, which is removed when the shoot is cut or grazed. The stems show varying degrees of lignification and are sometimes woody. Stolons, and less frequently rhizomes, are found in some species (e.g., white clover). The leaves may be simple or compound. The true clovers, for example, have trifoliate leaves which consist of two *stipules* at the base of the *petiole* and three *leaflets* (Figure 1-10).

Roots

The legumes are more deeply rooted than the grasses, having a well-developed main root, or *taproot*, with lateral branches. The way in which the plants develop root nodules in conjunction with bacteria from the genus *Rhizobium* is discussed in Chapter 6.

Inflorescence

The legume flowers, which are usually conspicuous, have a *calyx* with five *sepals* united at the base. The *corolla* has five petals, which are of three types: a single large *standard* or *banner petal*, two *wing petals*, and two united *keel petals* (Figure 1-11). With the keel petals are ten *stamens*, which are all united (or nine united and one free), forming a *staminal sheath* which encloses the *pistil*. An exception to this arrangement is found in the genus *Baptisia*, where the stamens are all free. The pistil is simple. The fruit is a *legume*, or pod, which opens along both the dorsal and ventral sutures and contains one to many seeds. The seed is normally without endosperm and has a thick seed coat or *testa*.

It is necessary for the pistil to be "sprung" or "tripped" by an insect, usually a bee, before pollination can take place. The bees are attracted to the flower by nectar which

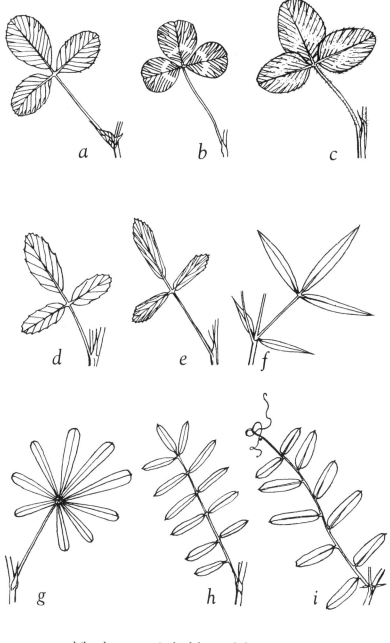

a. alsike clover
b. white clover
c. red clover
d. sweetclover
e. alfalfa
f. birdsfoot trefoil
g. white lupine
h. sainfoin
i. vetch

FIGURE 1-10
Compound leaves of forage legumes.

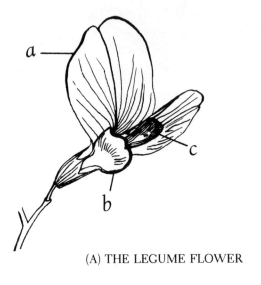

(A) THE LEGUME FLOWER

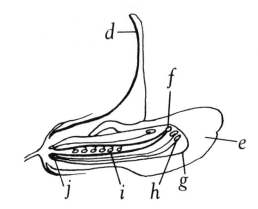

(B) CROSS SECTION OF THE FLOWER

a. banner petal
b. wing petals
c. keel petals
d. banner petal
e. wing petals
f. stigma
g. keel petal
h. anthers
i. ovaries
j. nectary
k. banner petal
l. staminal sheath
m. hooks retaining keel petals
n. wing petal
o. keel petal

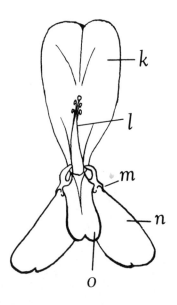

(C) DIAGRAM OF LEGUME FLOWER

FIGURE 1-11
Parts of the legume flower.

is secreted from glands at the base of the staminal sheath. Pollen is usually shed when the flower is in the bud stage. The stigma is not receptive until the flower is tripped, so cross-pollination usually results. Some species are entirely self-sterile, but where self-pollination is possible, it frequently leads to a loss of plant vigor, known as inbreeding depression. There are, however, some naturally self-pollinated legumes (e.g., sweetpeas).

FURTHER READING

Campbell, J. B., K. F. Best, and A. C. Budd, 1966, 99 Range Forage Plants of the Canadian Prairies, Agriculture Canada, Ottawa.

Gould, F. W., 1968, Grass Systematics, McGraw-Hill Book Company, New York.

Hanson, C. H., ed., 1972, Alfalfa Science and Technology, American Society of Agronomy, Madison, Wisconsin.

Hitchcock, A. S., 1951, Manual of the grasses of the United States, U.S.D.A. Misc. Publication 200.

Leithead, H. L., L. L. Yarlett, and T. N. Shiflet, 1971, 100 native forage grasses in 11 southern states, U.S.D.A. Handbook 389.

Nielsen, K. F., ed., 1969 Canadian Forage Crop Symposium, Western Co-operative Fertilizers Ltd., Calgary, Alberta.

Sprague, H. B., ed., 1974, Grasslands of the United States, Iowa State University Press, Ames, Iowa.

Tromanck, G. W., ed., 1969, Pasture and Range Plants, Phillips Petroleum Company, Bartlesville, Oklahoma.

chapter 2

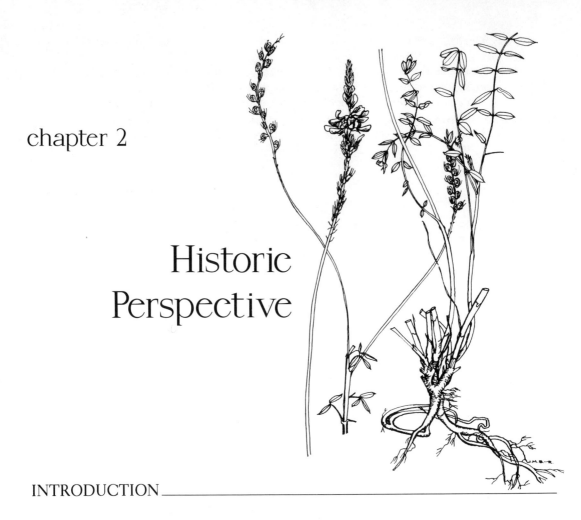

Historic Perspective

INTRODUCTION

As technology becomes more complex, it places an ever-increasing demand on our shrinking nonrenewable resources. As our world population expands, there is a growing pressure to increase agricultural productivity. A partial answer to both these problems lies in the more efficient use of our grassland areas. The grasslands of the world represent our greatest undeveloped, renewable natural resource. These regions are vast, environmentally diverse, and, more than any other part of the agriculture ecosystem, hold the promise of improved productivity in response to more intensive management methods. Also, forage production as a whole calls for less fossil fuel per unit of production than does any other type of farming.

GEOLOGY

Our grassland areas represent a natural resource which has existed for over 20 million years. On a global scale, they rank second only to forests in their total net primary production of biomass, and, by so doing, produce more than all other types of agricultural land. Thus, these areas, which long preceded the advent of people, can be regarded as a large proven renewable resource. From remotest antiquity, humanity has depended on forage species to support either the wild game hunted or domestic animals. The prairies and plains of North America, the pampas of South America, the veldts of Africa, and the steppes of Asia represent areas of climax vegetation determined initially by soil and climate. Such grasslands cover some 33 million km² (about 12.7 million square miles) and extend over 25% of the world's land surface.

Grasses and herbs appeared on the earth in the Tertiary era and diversified and spread during the mid-Tertiary period and the Miocene epoch (Table 2-1). They presented, then, a new environment which had not existed previously and which was exploited by herbivorous mammals. Global topography made these new regions rich in ecological habitats, and, in turn, evolution filled these niches with an ever-expanding and diverse range of grazing herbivores.

For example, in North America, the uplift which formed the Rocky Mountains caused changes in climate. Over a large area of the continent, dry summers, followed by drier winters, made the region unsuitable for the existing forests. Trees were replaced by grasses. Grasslands were developed not only in response to climate and soil, but in interaction with the animals that grazed them. Herds of buffalo and horses have cropped, trampled, and fertilized the prairies of North America. In parts of the world, this evolutionary association between plant and animal has existed for 30 million years. Both types of organism have acted as reciprocal influences on the other. In the grasses, hooks and hairs on the caryopsis, palea, and lemma facilitate seed distribution by animals. Also, many grasses are structurally well adapted to withstand grazing; short basal internodes, rhizome formation, branching by basal tillering, growth from a basal meristem, and lateral shoots from basal buds which replace the main shoot are all adaptations for survival under grazing. Also, as leaves are cropped, they are frequently replaced from below, while the sheathing leaves provide some protection for apical

TABLE 2-1
Geological time-scale in millions (10^6) of years

Era	Period	Epoch	Time before present 10^6 years
Cenozoic	Quaternary	Holocene	0.01
		Pleistocene	2.0
	Tertiary	Pliocene	7
		Miocene	22.5
		Oligocene	36
		Eocene	53.5
		Paleocene	65
Mesozoic	Cretaceous		135
	Jurassic		195
	Triassic		225

The upper and lower Palaeozoic eras follow.

meristems from trampling and grazing. A firm anchorage is provided by a fibrous root system in conjunction with underground stems. In most grasses, stem internodes elongate only when the inflorescence develops. The period when reproductive structures are vulnerable to the bite of a grazing animal is thus reduced to a minimum. Such was the evolutionary development of the 40 grass species used in our sown pastures today.

Of the grasses we use in pastures, none were derived from the American continent. By far, the largest number are species indigenous to the forest fringes of Eurasia and Africa. For this, too, there may well be a reason. The family Bovidae, to which our domestic cattle and sheep belong, originated in Eurasia in Pliocene times and moved into Africa during the late Pliocene period. Few species reached North America, and the grasses evolved there were developed in interaction with other types of grazing animals.

BEGINNING OF FARMING

After the evolutionary process had made possible the coexistence of grasses and grazing animals, the human species took its place in world history. Essentially, the grasses fill our needs in two ways. First, in the form of cereal crops, grasses provide grains which form the main carbohydrate in our diet. The advent of many great civilizations is associated with the availability of a productive carbohydrate crop. In the colder northern areas, oats and rye predominate, while wheat and barley are the most important crops in the temperate regions. Rice, maize, sorghum, and the millets feed the peoples of the tropical regions. Of these crops, rice and wheat are by far the most important. Second, the livestock which provide the proteins and fats in our diet are supported by forages and grasslands. The advance of knowledge and skill in growing cereal crops and in maintaining grazing animals has been closely linked with the advance and progress of civilization. This is still true today. Progress in our present civilization will depend on our ability to feed an ever-expanding world population.

The domestication of animals and the growing of cereal crops commenced some 12,000 years ago. In many areas, land was cleared for planting and game was hunted by burning the vegetation. There is evidence that this use of fire, combined with the use of grazing animals, brought about great changes in the world's vegetation. Regions previously dominated by trees and shrubs became areas of grassland and crops. As people settled in these areas, they became more dependent upon grasses for civilized life. However, if people were to adopt a settled way of life on the great plains of the temperate regions, they had to provide for their animals and themselves during the winter season. For their livestock, people had to preserve both straw from cereal crops and grass from the fields.

Alfalfa was the first cultivated forage plant. It was brought to Greece during the Persian invasion in 490 B.C. The Romans utilized it during their conquest of Greece and took it to Rome in 146 B.C. The Romans were the first to leave a written record of the many ways of making hay. Other writings show that clover was regarded as a valuable forage by the Saxons in 800 A.D. It was not, however, until the sixteenth century, that people started to sow grasses to form meadows for hay making and grazing. At this time, in Europe, rye grass *(Lolium perenne L.)* was used.

FARMING IN EUROPE

An even more important step was the appreciation of the value of mixing legumes with grasses to form a pasture. Just when this practice started is not certain, but Sir Richard Weston is usually given the credit for introducing red clover into pastures in England in 1613. By 1730, in England, Townshend had devised the four-course farming system which introduced the practice of alternating arable land with a period under pasture. The usual "ley," as the nonarable part of the rotation was called, was, at that time, red clover and ryegrass.

The next change in pasture production took place in the middle of the nineteenth century. This was the introduction of the use of fertilizers to improve growth. In the late 1880s, basic slag became available from steel industries. This phosphatic fertilizer, which was used in European farming in a general way, was also applied to pastures.

Application of artificial fertilizers calls for detailed scientific research if these substances are to be put to the best possible use. The world's first agricultural experiment station was established in 1843 at Rothamsted in England. While arable crops were given attention first, Lawes and Gilbert, working at Rothamsted, published, in 1882, a paper which gave details of the effect of fertilizer on forage production from permanent pastures. In 1889, experimental work on pastures commenced in Germany and spread to other parts of Europe.

In 1896, another important experiment was started at the Faculty of Agriculture Research Farm at Durham University, at Cockle Park, England. This experiment was different from that at Rothamsted in one important respect. The productivity of the plots was measured by the liveweight gains of sheep grazed on them and not by the weight of forage produced. These trials showed that basic slag aided the development of wild white clover in pastures, and that this pasture improvement was reflected in the liveweight gains of the animals. Further research at Cockle Park over the years indicated the most valuable combinations of clovers and grasses for British pastures. Thus, phosphatic fertilizers, white clovers, and the best grasses came to be regarded as important parts of pasture production.

It was not, however, until after the end of World War I that it was fully appreciated that pastures and grasslands should be regarded in much the same way as any other crop; that is, they called for the same attention from research workers and the same care and consideration from farmers. It was, in fact, food shortages (especially in the form of animal products) following the war which drew attention to this need. Investigations had been started in 1914 by J. B. Orr at the Rowell Research Institute at Aberdeen. He followed changes in substances of nutrient value in pastures through a 12-month cycle and investigated the effect on these of fertilizer applications and stocking rates. The most important of these new ventures was the establishment in 1919 of the Welsh Plant Breeding Station at Aberystwyth under the direction of R. G. Stapledon. Over the years, this station has provided a range of superior strains of pasture plants, as well as developing seed mixtures and advancing our knowledge of pasture management. Finally, in 1929, the Commonwealth Agricultural Bureau set up a department which acted in the clearing and exchange of information throughout the Commonwealth. In more recent years, the Food and Agriculture Organization of the United Nations has given additional impetus to the dissemination of knowledge concerning pasture and grasslands. Thus, pasture research started in the cool, humid, temperate regions of Britain and northwestern Europe.

FARMING IN NORTH AMERICA

In North America, developments followed another path. During the seventeenth century, the early colonists established themselves on the eastern side of the continent. Initially, they obtained their food by hunting, by gathering fruits, nuts, and berries, and by using the food crops of the American Indian peoples (beans, squashes, potatoes, peanuts, and corn). Since domesticated herbivores were not maintained by the Atlantic coast Indians, there were no local forage crops. The colonists used coarse, unpalatable local grasses, often from swamp areas, to feed the livestock which they had brought with them. As the livestock population increased, so did the settlers' interest in forage. Grasses were first introduced in 1670 and clover a few years later.

At that time, there were two types of land ownership. Either very large tracts of land were granted to an individual or to a trading company, or grants of about 50 acres were made to individuals. In all cases, there was little incentive to manage the land with long-term benefits in view. Changes in land tenure followed the Revolutionary War. Absentee landlords and "public" or "common" ownership of land were replaced by individual ownership. Settlers rushed to occupy the best areas before the land surveys were complete, and their claim to such land was recognized by the preemption law of 1841. After considerable deliberation, a bill giving free homesteads to individuals was passed in 1862. The combined outcome of these events was that the years 1790 to 1850 were a time of great expansion and migration.

Toward the end of the eighteenth century, a large number of agricultural societies were formed in the United States, the best known of which was the Philadelphia Society for the Promotion of Agriculture. These organizations encouraged the introduction of grasses and legumes from Europe, organized tests of the new species and cultivars, and issued recommendations as to the cultural methods and areas of use for the herbage crops.

As the pioneers moved westward, the forested areas, with their supply of wood for housing, fuel, and fencing and their supply of game for food were left behind. Some 290 million hectares (725 million acres) of prairie land stretched westward from Ohio. This vast area consisted of tall grasses which covered deep, fertile soils enriched over thousands of years by the decaying root systems of the native species. Even farther west, the six Great Plains states had extensive areas of short grass prairie.

Production from the new and fertile western farmlands, which grew small-grain crops and corn, caused a depression in eastern agriculture. Farmers in the east turned to the production of potatoes, vegetables, hay, orchard fruits, and dairy products. The development of railroads enabled the western states to provide eastern markets with wool, wheat, pork, and beef, while the industrialization of the east enabled that region to supply the western farmers with consumer goods. In the south, cotton became more and more widely grown to meet the demands of industrialized textile manufacture.

A most important influence in American agriculture was the development, between 1830 and 1860, of many new agricultural machines. The mechanical reaper, mechanical raker and binder attachments, the grain drill, and the threshing machine are but a few examples. Such new inventions were readily accepted on farms, for this was the time of the Civil War, when more than a million farmers and farm workers were absent with the army. Those who were left behind had a great need of extra help.

Following the Civil War, there was a period of vast agricultural expansion. As the industrialization of the country progressed, the ever-increasing city populations had to be fed, and the era of commercial farming began. This, in turn, increased the flow of people into the city. Farm mechanization continued, and fewer workers were needed on the farm. Extensive railroad systems and other improvements in transportation (including the car) led to small industries moving from the farm or village to the factory and city. The other event which contributed to economic development was the settlement of that part of the continent situated between the Missouri River and the Pacific Ocean. Between 1860 and 1900, the national farmland area was increased by more than 400 million acres. These new regions provided both a market for the industrialized east and more agriculture products to feed urban dwellers.

Unfortunately, much of this new area was very dry. Those who moved into such regions developed a livestock economy on ranges which were public domain. With his new and extensive ranching system, the American cowboy took his place in history. The early pioneers believed that the tall- and short-grass prairies would provide both arable and pasture land of virtually unlimited productivity. From Pennsylvania to Florida, large cattle herds moved west; from Texas and California, they moved north and east. From 1875 onward, there was an ever-increasing intensity in the pattern of land use, which terminated in land abuse. Experience drawn from the temperate and humid areas of the world was of no help in these drier prairie regions.

Even before 1890, it was evident that in Australia periodic droughts, overgrazing, and rabbits could, together, cause serious deterioration in grasslands. The treatment of the prairies and the Great Plains of America was to follow a similar pattern. Here burning and overstocking played a part. As the botanical composition of these areas changed, the more palatable and nutritive species were replaced by worthless types. This process culminated, after nearly a century of deterioration, with the droughts of the early 1930s.

It was evident that, where climax grasslands were situated in areas of low rainfall which experienced periodic droughts, careful management of the numbers and distribution of stock was essential. Adjustment of the stocking rate must be used to control the botanical composition of the pasture. Thus, grassland ecology and range management have been a fundamental part of grassland research in Canada, the United States, and Australia during the last 50 years. The problem of the management of pasture in temperate humid climates received attention in the second decade of this century. It was not until at least 10 years later that the problems of the drier grassland regions were recognized.

In the United States, as in Canada, the widespread droughts in 1933 and 1934 coincided with a time of economic depression. Farmers without an income from their crops were unable to purchase the goods whose production provided jobs for city workers. In response to this economic predicament, Congress passed, in 1933, the Agricultural Adjustment Act. This became the first of a series of laws enacted to assist agriculture. Subsequent legislation was aimed to alleviate problems caused by mortgage foreclosures and tax delinquencies. Studies of agricultural economics were undertaken. Land-use policies, aimed to offset the dangers of wind erosion, were implemented. Plans to provide credit for farmers were developed and agricultural research programs were started. For example, with the implementation of the Bankhead-Jones Act of 1935, which provided funds for scientific, technical, and economic research into farm problems, a total of over 800 new research projects were jointly initiated by state experiment stations and the U.S. Department of Agriculture.

In Canada, this new awareness of research needs resulted in the passing in 1935 of the Prairie Farm Rehabilitation Act. Funds became available for forage research on a much wider scale in western Canada. The Swift Current Experiment Station became a center of grassland survey and management research. Studies there first drew attention to the value of crested wheatgrass in providing a vegetative cover on abandoned or depleted farmlands. In Saskatoon, the establishment of the Forage Crops Laboratory in 1932 had an important impact on the breeding of crested wheatgrass, bromegrass, and alfalfa.

After World War II, the emphasis on forage research in Canada, as well as in the United States, again changed. Previously, forage had been important as a means of soil conservation and regeneration. Now production became of major importance. Consideration was given to intensifying grassland management rather than expanding the grazing area. Resistance to climatic stress and the production of more leafy swards were given consideration. Plant breeders produced new cultivars. This search for productive forage strains adapted to the farms and pastures of North America still continues.

In the last 80 years, changes on the farms have been very rapid. In 1905, farmers used seed and farmyard manure produced on their own farms, together with the labor of their own farm animals. Only about 25% of their input was purchased. Today, 75% of all farm input is purchased. This includes fertilizer, hybrid seeds, pesticides, petroleum, and a large and varied number of machines. With these new and powerful aids to production, farmers can manage large areas of land and individuals have increased their land holdings while other farmers moved off the land. The continuing exodus from the farm to the city has decreased only in recent years.

A consideration of past developments always leads to the contemplation of future prospects. How will forage production develop in the future? One thing is certain in farming today; high yields arising from the ability to absorb high inputs profitably are essential. Crops without this potential will not survive. With high-quantity forages, however, must go an improvement in quality. For any crop, the meaning of the word "quality" is difficult to define. For forages, quality means the ability of plants to produce material which contains nutrients in the proper proportions for a balanced ration. The production of strains of forages which not only give high yields but have the type of nutritional balance suitable for different classes of livestock presents an important challenge to plant breeders and forage crop managers in the years to come.

FURTHER READING

Arber, A., 1934, The Gramineae. A Study of Cereal, Bamboo and Grass, University Press, Cambridge.

Davies, J. G., 1951, Contributions to agricultural research in pastures, J. Aust. Inst. Agric. Sci. 17:54–66.

Frankel, O. H., 1954, Invasion and evolution of plants in Australia and New Zealand, Caryologia, 6 suppl., 600–19.

Hartley, W., and R. J. Williams, 1956, Centers of distribution of cultivated pasture grasses and their significance for plant introduction, Proc. Seventh Internat. Grassland Conf., pp. 190–9.

Shantz, H. L., 1954, The place of grasslands in the earth's cover of vegetation, Ecology 35:143–51.

Simpson, G. G., 1950, Meaning of Evolution—A Study of the History of Life and of Its Significance to Man, Yale University Press, New Haven, Connecticut.

Simpson, G. G., 1951, Horses, Oxford University Press, New York.

Stephens, C. G., and C. M. Donald, 1958, Australian soils and their responses to fertilizers, Advances in Agronomy 10:167–256.

Stewart, O. C. 1956, Fire as the first great force employed by Man. *In* Man's Role in Changing the Face of the Earth, University of Chicago Press, pp. 115–33.

Trumble, H. C., and A. B. Cashmore, 1954, The variety concept in relation to *Phalaris tuberosa* and allied forms, Herbage Reviews 2:1–4.

Wallace, L. R., 1959, Animal production from grassland—present problems and future needs, Aust. J. Sci. 21 (6a): 159–67.

Washburn, J. L., and V. Avis, 1958, Evolution and human behaviour. *In* Behaviour and Evolution, A. Roe and G. G. Simpson, eds., Yale University Press, New Haven, Connecticut, pp. 421–36.

Weaver, J. E., 1954, North American Prairie, Johnson Publishing Co., Chicago.

chapter 3

The Grasses

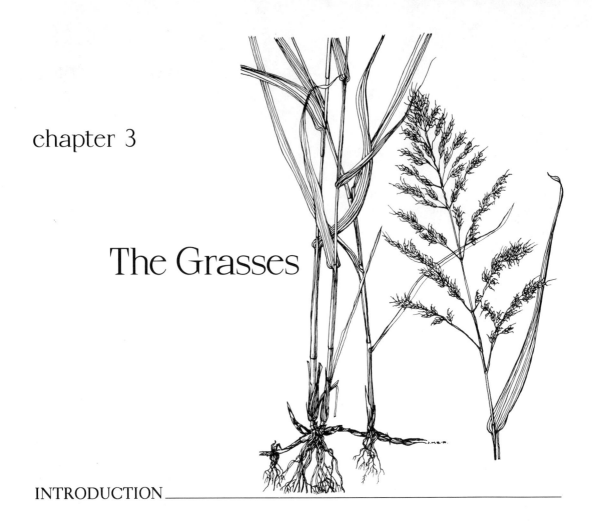

INTRODUCTION

The grasses have been classified in very many ways. Most of these classifications are based on morphological characteristics which frequently cannot be related to cytological, genetic, or physiological plant characteristics. In this chapter, we will consider the cultivated grasses in relation to their area of adaptation and their agricultural use. On this basis, cultivated species fall into two large groups: the cool-season grasses and the warm-season or tropical grasses. In this chapter, we will discuss only perennial species. The annual species will be considered in Chapter 5.

Considering the North American continent as a whole, introduced forage grasses are grown throughout the east up to a line which runs north and south in the western parts of North Dakota, South Dakota, Nebraska, Kansas, Oklahoma, and Texas. West of that line, native short grasses are used. Farther west again, native desert grasses and shrubs provide grazing for livestock. On the west coast of the United States and in Canada, introduced northern grasses are cultivated, except on the west coast of California, where winter annual species are grown. A line drawn from east to west, and passing through North Carolina, Tennessee, and Arkansas, separates the areas in which cool- and warm-season perennial grasses are cultivated.

COOL-SEASON PERENNIAL GRASSES

Most of the cool-season species originated in northern Europe and in the northeastern parts of Asia. They grow best in a cool, moist climate and are usually winter hardy. Not all these grasses are particularly drought tolerant, but some species, such as smooth bromegrass, are able to withstand dry conditions. Their requirements for floral induction are often complex. There is frequently a need for long days, but this may be preceded by short-day and low-temperature requirements (see Chapter 15). When grown for hay, plants of this type usually produce a large first cut, followed by a smaller aftermath. The main carboxylating enzyme for these grasses is ribulose-1,5-diphosphate carboxylase, which is present in the C_3 (Calvin) photosynthetic pathway. The seed sizes for the various species discussed in the following pages are indicated in Table 6-4.

Bentgrasses (genus *Agrostis*)

Members of this genus are commonly grown in Britain and continental Europe, but are little used in the United States, except as turf grasses. The genus contains several species of agricultural value, such as colonial bent *(Agrostis tenuis)*, creeping bent *(A. palustris)*, velvet bent *(A. canina)*, and the most widely used species, redtop *(A. alba)*. The bentgrasses are in general somewhat lower yielding, have poorer quality, and are less palatable than other grass species. These species are able to tolerate poor soil conditions, possibly because they have dense shallow roots which keep the plants supplied with nutrients. They have a creeping habit and are persistent once established. Redtop (Figure 3-1) will grow under a wider range of conditions than any other grass and does especially well where there is an abundant supply of water which might even be excessive for other grasses.

Production from the bentgrasses is much improved by growing them in a mixture with a legume. The chemical analysis of such mixtures shows a good nutrient balance for animal feed. Bentgrasses are excellent soil builders. Their extensive root system increases soil organic matter content. Consequently, they are widely used under poor hill-farm conditions in Britain. Bentgrasses also produce a turf suitable for golf courses. Velvet bentgrass is the species most suitable for putting greens and areas where a very fine turf is required in the northern United States. Colonial bentgrass is a grass most commonly used in lawns, being shorter than redtop.

Penncross, a well-known and widely used cultivar of creeping bentgrass, originated from clones selected in Philadelphia.

Bluegrasses (genus *Poa*)

Members of this genus, which are known as *meadow grasses* outside the American continent, are low to moderately tall in habit. All the species which are of agricultural interest have folded leaves, a spreading panicle, spikelets which are small and flattened, keeled lemmas, no awn, and are mainly perennial. Of approximately 250 species which

FIGURE 3-1
Agrostis alba. Redtop.

make up the genus, 59 are found in North America. The genus has a basic chromosome number of 7, but much of the reproduction is *apomictic*. The species used most commonly in agriculture are Canada bluegrass (*Poa compressa* L.), early bluegrass (*P. cusickii* Vasey), big bluegrass (*P. ampla* Merr.), roughstalked bluegrass or meadow grass (*P. trivialis* L.), mutton bluegrass (*P. fendleriana* Vasey), and Kentucky bluegrass or smooth-stalked meadow grass (*P. pratensis* L.). Although many members of the genus are native to North America, Canada bluegrass, rough-stalked bluegrass, and Kentucky bluegrass, the most widely used members of this genus, were introduced from Eurasia.

Kentucky bluegrass, also called Kentucky blue, bluegrass, Junegrass, and smooth-stalked meadow grass, was first introduced into North America from Europe by settlers in about 1625 (Figure 3-2). There appears to have been more than one introduction. The French traders were responsible for carrying the species south and west to the Ohio and the Mississippi rivers and to areas near the Great Lakes. The species established well in its new environment and spread so rapidly that it was called "white man's foot grass" by the North American Indians. It grows well in association with white clover, which was also called "white man's foot." Kentucky bluegrass is now widely found in both cultivated and native pastures in North America, Europe, and northern Asia.

Kentucky bluegrass is easily identified in the vegetative stage by its leaf tips, which are shaped like the bow of a boat. Another diagnostic characteristic is the web of white hairs at the base of the "seed" when the plant is in head. Kentucky bluegrass is well adapted to cool, moist climates (over 500 mm rainfall) where soils are relatively fertile and have a pH between 6.0 and 7.0. These conditions prevail in Kentucky and much of the northeastern parts of the United States, where the species is highly productive and very long lived. Being of low stature, it is especially well suited to grazing and, like white clover, is able to withstand frequent clipping. Cattle find young shoots of the species very palatable. Its size makes it an unsatisfactory hay species, and it is somewhat unpalatable when mature. It spreads readily and develops a thick sod. Kentucky bluegrass is relatively shallow rooted, with about 80% of the root system in the top 15 cm of the soil. As a result, it is especially susceptible to periods of drought during the summer months; 2 weeks without rain will reduce yields substantially. For this reason, yields of Kentucky bluegrass are frequently lower than those from species with deeper root systems, but the spring growth is palatable and nutritious. Kentucky bluegrass does not persist well under heavy shade, but is frequently the first species to appear, together with white clover, when forested areas are cleared.

Kentucky bluegrass responds very well to close defoliation by developing a dense vegetative cover and producing higher yields than when it is not cut (Table 3-1). While the total annual yield of Kentucky bluegrass may be low, especially where its climatic and soil requirements are not met, the spring growth is very good and yields at that time are high. Its crude protein content is slightly lower and acid-detergent fiber content a little higher than those for orchardgrass and timothy.

Considering its wide distribution, relatively few varieties of Kentucky bluegrass have been developed for agriculture use. *Merion* is used in both the United States and Canada, where *Kenon* is also grown. The cultivar *Delta* is widely grown in the United States. Because of its ability to withstand repeated defoliation, this species is frequently used as a lawn grass, in athletic fields, in parks, and beside roads and waterways. For

FIGURE 3-2
Poa pratensis L. Kentucky bluegrass.

TABLE 3-1
Number of shoots and rhizomes per square meter
(square feet in parentheses) of Kentucky bluegrass turf

Plant part	No fertilizer		Nitrogen applied	
	Not mowed	Mowed	Not mowed	Mowed
Shoots	38.6 (3.5)	77.4 (7.0)	69.8 (6.3)	126.5 (11.4)
Rhizomes	23.0 (2.1)	30.0 (2.7)	39.9 (3.6)	38.3 (3.5)

(Source: Ohio Agr. Expt. Sta. Res. Bul. 681)

these purposes, it has received attention from plant breeders, who have selected dwarf strains which will withstand very close cutting.

Smooth Bromegrass (*Bromus inermis* L.)

Smooth bromegrass is one of the most extensively grown species in the northern parts of the North American continent. There are two types of this grass, which is known variously as bromegrass, Austrian brome, Hungarian brome, Russian brome, and smooth bromegrass. The southern type was introduced into the United States from France and Hungary and was subsequently used in Canada. The northern type was first introduced into Canada from Germany in 1888 and is now extensively grown on the Canadian prairies. The northern strains give higher seed yields than do the southern types when grown in Canada, while the reverse is true for the northern United States. The northern types produce a more open sward with slightly shorter plants than do the southern strains. They are adapted to, and preferred in, Canada and in the northwestern United States. The southern types, which produce a sod-forming plant, are well adapted to the Corn Belt of the United States.

The genus *Bromus*, a member of the tribe *Festuceae*, consists of some 60 species, both introduced and native to North America. Smooth bromegrass belongs to the subgenus *Zerna*. The name *Bromus*, which is Greek for oat, indicates the open nature of the panicle, which has a floret with a thin, rounded, and awnless lemma (Figure 3-3). *Bromus inermis* is a rhizomatous, long-lived, perennial which is adapted to most temperate climates. It is resistant to drought and to extremes of temperature, being capable of withstanding both hot, dry summers and long, cold winters. The species is grown both alone and in mixtures with other grasses and legumes, and is used for pasture, hay, and erosion control. The forage quality of smooth bromegrass ranks well among the cool-season grasses, and it is more palatable than most species in the vegetative stage. The crude protein content is high, ranging from 12% to over 20%, during the time of rapid growth at the beginning of the season. The inflorescence consists of many spikelets, each of which contains several hermaphroditic florets. Each panicle usually produces abundant seed and is highly cross-fertilizing, naturally cross-pollinated, and rather self-sterile (Figure 3-4).

Following the introduction of smooth bromegrass into the North American conti-

FIGURE 3-3
Bromus inermis L. Smooth bromegrass.

FIGURE 3-4
The panicle of smooth bromegrass is frequently one sided as the plant approaches maturity.

nent and the initial recognition of the value of this species, interest declined. It was not until farmers and scientists sought to combat the dustbowl conditions generated in the early 1930s by a combination of extensive plowing, overgrazing, and drought conditions that it was realized that smooth bromegrass was, among the introduced grasses, one of the principal and most widespread survivors. The species is now accepted as one of the most successful grasses used for erosion control on roadside shoulders and steep road cuts. It is frequently grown in mixtures with other grasses for the vegetation of waterways, irrigation canal banks, and terraces, as well as in areas where the soil has been extensively disturbed. The extensive root system of this grass species brings about rapid improvement in soil structure. Where a legume has been incorporated, the decay of the legume root maintains a balance of available nitrogen and aids in the decomposition of the grass roots. In areas where subsoil has been exposed or where soil has been eroded on slopes, the establishment of the species is much enhanced by a dressing of a nitrogenous fertilizer.

On the Canadian prairies, smooth bromegrass is one of the principal grass components of a grass–alfalfa seed mixture which is widely used on nonirrigated pastures. The most favorable time for the establishment of smooth bromegrass is in the early spring. Planting should be carried out as early as possible. Where pastures are used for grazing only, rotational grazing will double the production of beef per hectare (acre) over that obtained from continuous grazing (see Chapter 17). Like many other tall grasses, the yield of smooth bromegrass decreases with frequent cutting. Four defoliations, under grazing conditions, and two cuts where the material is intended for hay give optimum productivity and a satisfactory balance between production and forage quality. For both hay and pasture, smooth bromegrass behaves satisfactorily, if not ideally, in a mixture with alfalfa. *Baylor* and *Saratoga* are two widely used cultivars.

Fescues (genus *Festuca*)

The inflorescence of species of the genus *Festuca* is a panicle with paired branches of unequal size. All members of the genus are perennials. Annual species with morphological characteristics similar to the fescues are usually placed in the genus *Vulpia*, but are sometimes included in the genus *Festuca*. Those fescues which are of agricultural importance fall into two groups. First are the species with broad, flat leaves, of which meadow fescue (*Festuca elatior* L.) and tall fescue (*F. arundinaceae* Schreb.) are the best-known examples. In the second group, all species have narrow, stiff, pointed leaves which are permanently folded. The species from this group in common agricultural use are red fescue (*F. rubra* L.), sheep fescue (*F. ovina* L.), rough fescue (*F. scabrella* T.), and bluebunch or Idaho fescue (*F. idahoensis* Elmer).

Tall fescue is often confused with meadow fescue. Until 1950, the two species were frequently regarded as one. Both species were introduced to North America from western Europe. The initial confusion concerning their names, together with their morphological similarity, makes it difficult to interpret the data from tests conducted toward the end of the nineteenth century. Today, tall fescue is regarded as a species with a very wide range of adaptation. It is grown, on a world scale, from Siberia in the north to the hill areas of the Malagasy Republic and South Africa in the south. On the North American continent, tall fescue is found from Florida in the United States to the Peace River area of Alberta, Canada (Figure 3-5).

The species grows well under both acid (pH 4.7) and alkaline (pH 9.5) soil conditions, and its lower rainfall limit appears to be about 45 cm (18 in.) per year. Under these conditions, even on poor soils, it will produce satisfactory growth. However, it does especially well under moist conditions, for it can withstand short-term waterlogging and high moisture levels even on heavy soils. It also gives excellent erosion control and is valuable for the re-establishment of vegetation in disturbed areas.

High temperatures (over 25°C or 77°F) have a less detrimental effect on the growth of tall fescue than does inadequate soil moisture. Thus, if irrigation is available, this species could well be used as far south as Arizona. It becomes dormant under conditions where there is both temperature and water stress. While tall fescue has a very wide adaptation, it is not necessarily the most productive species over the whole area. Its area of major use in the United States is in Virginia, Kentucky, Missouri, the Carolinas, Tennessee, Arkansas, and the northern parts of Mississippi, Alabama, and Georgia. It is also grown outside these regions, in Oregon.

Tall fescue, when fed to beef cattle in Alabama, gives average daily weight gains which compare favorably with those obtained with reed canarygrass and orchardgrass (Table 3-2). Animal weight gains per hectare (acre) are also high. Tall fescue contains alkaloids of the loline family, as well as perloline and perlolidine. The latter two substances have been shown to be mildly toxic to cattle. Although the tall fescue sward

FIGURE 3-5
Festuca arundinaceae Schreb. Tall fescue.

TABLE 3-2
Average annual performance of steers on tall fescue and phalaris pastures in west central Alabama, 3-year means

	Stocking rate	Beef gain	Average daily gain
	steers/ha	kg/ha	
Tall fescue	3.53	488	0.81
Phalaris	3.15	389	0.78
	steers/acre	lb/acre	
Tall fescue	1.41	434	0.72
Phalaris	1.26	346	0.69

(Source: Hoveland et al., 1980, Agron. J. 72. By permission of The American Society of Agronomy, Inc.)

has the open appearance normally associated with a bunchgrass, this species has very short rhizomes. The culms and leaf sheaths are smooth, but the shiny, dark green, ribbed leaves are barbed at the margins. The plant can reach heights of 2 m (6 ½ ft), when in head, with leaves 60 cm (nearly 2 ft) in length. It is a long-lived plant with a deep, tough, and extensive root system, and it will survive trampling and compaction better than most species.

The first cultivar to be developed in the United States was *Alta*, which was selected in 1918 in Oregon. *Kentucky 31*, collected in Kentucky in 1931, is another well-known cultivar which was widely grown and still in use today.

In recent years *Fawn* has out yielded both of these cultivars in trials conducted in Oregon; giving about 16% more dry matter and 25% more seed. In 1965 the Kentucky Agricultural Station released a cultivar called *Kenwell* which also out yields Kentucky 31. *Goar* is from one to two weeks earlier than Kentucky 31 and *Missouri 69*, released in 1977, and is reported to give from 35% to 40% more daily gain when fed to steers than Kentucky 31. *Kenhy* is a special interest cultivar since it is derived from a ryegrass-tall fescue cross. This cultivar is superior in quality to Kentucky 31, having a low level of perloline (see Chapter 10).

Meadow fescue, a perennial bunchgrass from Europe, differs from tall fescue ($2n = 42$) in that it has 14 chromosomes ($2n$). In general, meadow fescue is a smaller plant and can be morphologically distinguished from tall fescue by its long, curved, glabrous auricles (Figure 3-6). Also, meadow fescue has a panicle which usually has paired branches of unequal size, some with only one or two spikelets. In contrast, the tall fescue panicle has three or more spikelets on all branches.

Meadow fescue is more winter hardy than tall fescue and is consequently used more widely in the north and in the east. It gives high yields under irrigation in both southern Alberta and Saskatchewan. On the edge of prairies, it gives satisfactory yields in

FIGURE 3-6
Festuca elatior L. Meadow fescue.

mixtures with brome or timothy. It is, however, less palatable and shorter lived than either of these two species.

Red fescue is a variable species of which two forms are of special importance: *F. rubra* sub-sp. *commutata* (Chewing's fescue) from New Zealand, with 42 chromosomes, and *F. rubra* sub-sp. *rubra* (creeping red fescue) with 56 chromosomes (both $2n$ values). Both have small, pointed, folded leaves which are bright green in color. Short brown rhizomes enable the creeping red fescue plant to spread over an area of several meters. Chewing's fescue, on the other hand, is noncreeping. The species is found naturally in North Africa, Europe, Asia, and North America, where it is present in the Rocky and Allegheny mountains (Figure 3-7).

Creeping red fescue is widely used as a "bottom" grass. Being a small creeping plant, it is well able to fill in the lower part of the pasture canopy. It is less heavily grazed during the summer, when tall species are available. In the fall, it retains a high protein content (about 15% crude protein) when other species have become fibrous. Creeping red fescue also grows farther north than most other cultivated grasses. It does especially well under the dry conditions found in the foothills of the Rocky Mountains in Alberta and will grow better on poor soils than either timothy or Kentucky bluegrass. It yields well in mixtures with alfalfa, white or alsike clover, and bromegrass. Because creeping red fescue develops a dense turf and tolerates trampling, it is valuable for erosion control. It is quite unsuitable for hay, having a fine leaf as well as a low growth habit.

Chewing's fescue has a lower growth habit than creeping red fescue and is consequently suitable for lawns and golf fairways. A typical cultivar of this type is *Highlight*, which was developed in Holland and is now widely used in North America. Commonly used cultivars of creeping red fescue are *Boreal* and *Reptans*. *Reptans* is a Swedish cultivar. *Dawson*, a cultivar bred in England, is intermediate in height between the creeping red fescues and Chewing's fescues.

Meadow Foxtail (*Alopecurus pratensis* L.)

This species has been grown in Europe and Asia for about 250 years. Its main use is for long-term pasture and, to a much lesser extent, for hay and silage. The species is not especially sensitive to temperature, but yields best under cool, moist conditions. In North America, it is grown in the Pacific Northwest and to a limited extent in northern Canada and central Alaska. No doubt, its full potential has been overlooked in these latter two areas (Figure 3-8).

Meadow foxtail is well able to withstand cold weather and frost during the growing season and can survive long, cold winters. Provided that soil is reasonably fertile, it will survive in wet conditions and withstand some flooding. It is also tolerant of wet alkaline soils (pH 8.0 to 8.5) and will resist such conditions well, once it is established. It is not drought tolerant and quickly dies out where soils are dry. Meadow foxtail gives good early growth, and with the exception of crested wheatgrass, produces grazing before all other grasses. It is palatable and is eaten with relish by all classes of stock. The protein content is as high as, or slightly higher than, that found in timothy (18% crude

FIGURE 3-7
Festuca rubra L. subsp. *rubra*. Creeping red fescue.

FIGURE 3-8
Alopecurus pratensis L. Meadow foxtail.

protein), while its fiber content is low. When the seed is mature, the whole of the single-seeded spikelet sheds readily from the inflorescence. This leads to two problems. First, the seed cannot be harvested without substantial losses and, second, the seed is "fluffy" and difficult to clean. From a pasture point of view, this is an advantage, since, once established, the species is self-seeding. Seed production, however, is a difficult and costly process, and the seed is frequently of low germination and poor purity. Also, since the meadow foxtail seed is light, it does not flow well in the seed drill, and this adds to establishment problems.

Orchardgrass (*Dactylis glomerata* L.)

There are reports from as early as the fifteenth century of the use of this species in pastures in Europe (where it is called cocksfoot). It was introduced to North America in colonial times (about 1750), but did not become popular until 1940. At that time, ladino clover, which grows well with orchardgrass, came into general use, and the two species were grown widely together.

Orchardgrass is a bunch-type grass with neither rhizomes nor stolons. It can be used for pasture, hay, or silage, but since it becomes coarse as it matures, it is most commonly used for pasture. The thick, fleshy base of the leaf sheaths stores reserves of plant "food" material. Where grazing is light, this reserve enables the plant to grow vigorously, and individual plants may form large, coarse and unpalatable clumps in the pasture. The species grows vigorously early in the grazing season, and high stocking rates are an important feature of management at that time in order to prevent the formation of these unpalatable clumps. While it is possible to overstock by using close, continuous grazing, such problems may be avoided by rotational grazing (see Chapter 17).

Orchardgrass responds well to nitrogen fertilizers. The literature is full of examples in which the addition of nitrogen has increased yields from two- to fourfold. Adequate soil moisture is essential for high returns from fertilizer applications. Orchardgrass yield may be increased by irrigating. Increases of from 50% to 100% have been reported from Pennsylvania and from the Pacific coast and intermountain states, when irrigation water is available.

Where a legume is present in a pasture mixture, there is a danger that nitrogen, under conditions of plentiful soil moisture, will result in orchardgrass suppressing the legume. Orchardgrass has a deep, vigorous root system which enables it to compete with a legume. Because the grass is also able to shade and so retard the growth of the legume, it is sometimes difficult to maintain legumes in the sward. This problem is most marked with ladino clover, but is also noticeable with alfalfa.

Orchardgrass is well able to withstand shading and is frequently grown in orchards; hence, its name. It is neither as winter hardy nor as drought tolerant as smooth bromegrass. However, where soil moisture and nutrients are adequate, and where selected strains are used, the species has been shown to be long lived as far north as Edmonton, Alberta, and in the Piedmont region of the Carolinas.

Orchardgrass leaves are folded in the bud, broad and dull gray to blue-green in color. The tillers are flat. The ligules are large, white, and conspicuous. There are no

FIGURE 3-9
Dactylis glomerata L. Orchardgrass.

TABLE 3-3
Proximate feeding stuff analysis results for orchardgrass and tall fescue grown in Virginia

	Tall fescue %	Orchardgrass %
Dry matter	29.6	33.6
Crude protein	9.9	9.3
Ether extract	2.2	2.6
Nitrogen-free extractive	48.8	45.7
Acid-detergent fiber	41.3	44.6
Acid-detergent lignin	6.9	7.8

(Source: Rayburn et al., 1980, Agron. J. 72)

auricles. The inflorescence is a large panicle, with long primary branches and short ultimate branches which give the head its characteristic appearance (Figure 3-9). The spikelets are borne in dense clumps; hence, the Latin name *glomerata*, which means "clumped." Consequently, it is sometimes confused with reed canarygrass. Orchardgrass species compares well with tall fescue for quality characteristics (Table 3-3).

There is a very wide range of orchardgrass cultivars available. *Avon* is a high-yielding cultivar which has good winter hardiness. *Chinook* was bred at Lethbridge in western Canada, but gives rather lower yields than Avon. *Kay* has frequently given higher yields than Avon. *Sterling*, an orchardgrass cultivar produced in Iowa, gives good establishment and high yields, but it is reported to be susceptible to rust and to leaf streak.

Reed Canarygrass (*Phalaris arundinaceae* L.)

This grass is a native of Europe and Asia. Its first agricultural use was reported from Sweden in 1749. It was used in England in the early nineteenth century and had been grown in France and Germany by 1850. The first known seeding in North America was about 1900. This species is now grown from coast to coast in both the northern United States and in Canada. There are large areas established on the west coast and in the humid north-central regions.

Reed canarygrass will tolerate low lying, poorly drained soils. It spreads aggressively by means of well-developed scaly rhizomes and forms a dense sod. Yields are relatively high. It becomes tall and coarse as it matures and, if grazed, it must be kept from reaching maturity. A pasture height of about 30 cm (12 in.) is most satisfactory. The pasture should be permitted to make a good start in the spring and thereafter be continuously grazed. The rate of regrowth is normally rapid and uniform.

This species is palatable to beef cattle and horses, but unpalatable to sheep. The protein content of the reed canarygrass herbage is not high under cool conditions (about 11% crude protein); this can be increased to 13% crude protein by nitrogenous

FIGURE 3-10
Phalaris arundinaceae L. Reed canarygrass.

fertilizer applications. Reed canarygrass gives acid-detergent fiber values (Chapter 9) lower than those of smooth bromegrass, and its digestibility has been reported to be equal to that of alfalfa. In spite of its apparent good nutritive value, poor performance has been reported in many places (Michigan, Minnesota, British Columbia, and Germany) for all classes of livestock. Low intake and lack of palatability are associated with mildly toxic alkaloids found in reed canarygrass. This problem will be considered in Chapter 10, where antiquality characters will be discussed.

Reed canarygrass has been successfully used to stabilize banks or waterways and for gully control. Rhizome pieces planted in such situations quickly develop into a firm sod, growing through silt deposits and withstanding water flow. The root system is very much larger than that of bromegrass, producing 19,000 kg of roots/hectare (16,910 lb/acre), as compared with 12,300 kg/ha (10,947 lb/acre) for smooth bromegrass. Figure 3-10 shows reed canarygrass to be a tall, coarse, perennial species which produces a panicle that is initially compact and cylindrical, but which opens up at anthesis.

Castor is a cultivar of reed canarygrass bred at Beaverlodge for its high seed-retaining character (which facilitates seed harvest). The seed yield of Castor is twice that of *Frontier*, an older variety from Ottawa. Plant breeding work with this species is now being directed to produce low-alkaloid strains with improved palatability, has led to the release of *Vantage* which has a low-alkaloid content.

Ryegrasses (genus *Lolium*)

The common name for this genus is not appropriate, since these plants, which are members of the tribe *Festuceae*, in no way resemble rye. Their spikelets are many flowered and not completely enclosed by glumes as is typical for their tribe. They differ from other members of the *Festuceae*, however, in that their spikelets are borne on a spike rather than a panicle. Also, in the genus *Lolium*, the spikelet is held with its edge toward the rachis, not at right angles to it, as is normal in other genera (e.g., wheat). For a normal spikelet, such an arrangement would result in one of the glumes being adjacent to the rachis. In the case of *Lolium*, this glume is absent and the rachis internodes are hollowed. The last spikelet, at the apex of the spike, has two glumes.

There are two members of the genus which are of considerable agricultural importance on a world scale. They are perennial ryegrass (*Lolium perenne* L.) and annual or Italian ryegrass (*L. multiflorum* Lam.), which is an annual. Both species are native to Europe and north Africa and have been in agricultural use since 1580 or earlier.

Perennial ryegrass is a short-lived bunch grass with a shallow root system. The leaves are folded in the bud and the whole plant is glabrous. The auricles are very small; the ligule is blunt and short. The leaf sheath is split to the base, where it is a cherry-red color (Figure 3-11).

The plant is nutritious and palatable and stands up to hard grazing. It will not do well under poor conditions, where fertility or rainfall is low. It requires an annual rainfall of between 850 and 1,030 mm (34 and 41 in.) and a mild climate during the growing season. However, it has the ability to survive long, cold winters. The species normally establishes easily. Being of somewhat small stature, it cannot be made easily into hay. Perennial

FIGURE 3-11
Lolium perenne L. Perennial ryegrass.

ryegrass is a common pasture grass in western Europe, New Zealand, and the northeastern and northern United States. Many cultivars and a number of hybrids, both diploid and tetraploid, are in use. *Norlea,* a cultivar developed in Ottawa, is commonly used in eastern Canada and the United States, while *Pacific* is favored in the west.

Annual ryegrass is an annual or short-lived perennial. The inflorescence of this species closely resembles that of perennial ryegrass, but normally has more florets per spikelet and a fine terminal awn. Until the early nineteenth century, it was relatively unknown, being used only in northern Italy. Now it is recognized as a very quick growing, high-yielding species which establishes readily. It is a valuable component of short-term pastures and is used mainly for grazing (Figure 3-12).

Wildrye (genus *Elymus*)

The spikelets in the genus *Elymus* are rather like those of *Agropyron,* but differ in that they have two spikelets side by side at each node of the rachis (rather than only one, as in *Agropyron*). There are a number of species in this genus which are native to North America (e.g., Canada wildrye), but the two species which are most commonly cultivated have both been introduced. Altai wildrye (*E. angustus* Trin.) comes from Mongolia, and Russian wildrye (*E. juncea* L.) grows wild on the steppes of Siberia and central Asia.

Canada wildrye (*Elymus canadensis* L.) is one of the few North American native species which is cultivated. It is a large, coarse, perennial bunchgrass which is found throughout the United States and Canada, but is most common in the northern Great Plains and the Prairie Provinces.

The species yields well under relatively cool conditions on moist soils. It produces fairly good quality hay if harvested just before the head emerges from the boot. It is a palatable pasture grass which grows well in a mixture with crested wheatgrass. It starts growth rather later in the spring than either smooth bromegrass or crested wheatgrass. The young seedlings are extremely vigorous. The seedheads are blue-green and turn dark purple when mature. The leaves are broad, flat, and rough. *Mandan* wildrye is a cultivar selected for high forage and seed yield in North Dakota.

Russian wildrye was first introduced into North America from Siberia by the University of Saskatchewan in 1926, and was subsequently used to combat the dustbowl conditions of the early 1930s. It was found to be an excellent pasture grass for the dry southern parts of the Canadian prairies and for the adjacent central and northern areas of the United States (e.g., Montana and the Dakotas). Russian wildrye is an extremely long-lived perennial bunchgrass with an extensive fibrous root system (Figure 3-13). The seedling is small and grows very slowly, making the species difficult to establish. This is especially true on sandy soils in dry areas, since the root system may not develop to an adequate depth before the soil surface dries out. It is not a high-yielding grass in terms of either seed or forage production. Once established, however, it is exceedingly hardy and has marked tolerance to both drought and cold. The species

FIGURE 3-12
Lolium multiflorum Lam. Annual or Italian ryegrass.

FIGURE 3-13
Elymus juncea L. Russian wildrye.

will also withstand saline conditions. It has a long growing period, starting quite early in the spring and continuing late into the fall.

The protein content is high throughout the season, but for both protein content and palatability, it compares especially well with other species in the fall. From August to November, it can provide highly digestible forage which is relished by livestock. Breaking and reseeding native prairie with Russian wildrye and alfalfa provide excellent possibilities for increased pasture production.

The plant produces a large number of dark green basal leaves which have large ligules, but no auricles. The spike is borne on long (about 1 m or over 3 ft), nearly leafless culms. Given these characteristics, the plant is obviously unsatisfactory for hay making, since the basal clumps of leaf are hard to cut. Hence, its use is primarily for pasture. The dense spike shatters readily, making seed harvest difficult.

Altai wildrye was introduced to the Agriculture Canada Research Station in Ottawa in 1934. It was evaluated in western Canada toward the end of the 1970s, at which time selections were made which led to the release of improved cultivars. No doubt, this long delay between introduction and development for commercial use resulted from a failure to realize that this species has some advantages over Russian wildrye, which usually outyields Altai wildrye (Figure 3-14). In fact, the two species have much in common, both being winter-hardy, saline- (Figure 3-15) and drought-tolerant, long-lived perennial pasture grasses. Altai wildrye, however, is especially valuable for grazing from October to early February. At that time the digestible nutrients for this species are higher than those found in any other cultivated or native grass (Figure 3-16). Also, the basal leaves are erect and frequently project above the snow cover, thus making winter grazing possible. The plant is slightly rhizomatous but has a bunched appearance in the field. The head is an open spike and the culm is nearly naked. It establishes easily and emerges from greater seeding depth than any other species (Table 3-4), an important advantage under dryland farming conditions. The plant is also very deep rooted, penetrating the soil to a depth of 3 to 4 m (3.3 to 4.4 yd). There is only one variety, *Prairieland*. It was bred by T. Lawrence at the Agriculture Canada Research Station at Swift Current.

TABLE 3-4
Percentage of emergence of four grasses from four depths of seeding in the greenhouse

GRASS SPECIES	DEPTH OF SEEDING			
	3 cm (1 in.)	5 cm (1.8 in.)	7 cm (2.5 in.)	9 cm (3.25 in.)
Altai wild ryegrass	85	75	58	40
Russian wild ryegrass	78	54	10	2
Bromegrass	89	73	33	5
Tall wheatgrass	86	57	15	3

FIGURE 3-14
Elymus angustus Trin. Altai wildrye.

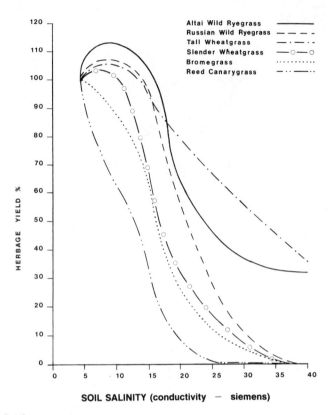

FIGURE 3-15
Herbage yield of six grasses at varying levels of soil salinity as a percentage of the yield of each grass on a nonsaline soil. (Source: T. Lawrence, 1977, Agriculture Canada Publication 1602)

Timothy (*Phleum pratense* L.)

This species is a hexaploid (2n = 42). Cytological studies, conducted in 1930, show that timothy may be synthesized by crossing a plant formerly known as *Phleum nodosum*, now called *P. bertolonii* D.C. (2n = 14), with *P. commutatum* Grand (2n = 28). The progeny of this cross is a sterile triploid hybrid from which the hexaploid timothy plant may be obtained by chromosome doubling. Where and when this process took place in nature, so that *P. pratense* might evolve, is uncertain, but there is evidence that it occurred in relatively recent times. The tetraploid species (*P. commutatum*) is an alpine plant which, in central Europe and in Britain, is found only at high elevations. Opportunities for crossing with *P. bertolonii*, the species used in European agriculture in the 1600s, would appear to have been few. However, very early in the settlement of the North American continent, *P. bertolonii* was introduced by the immigrants. This species spread rapidly throughout the New England states. There, *P. commutatum* is

Cool-Season Perennial Grasses / 51

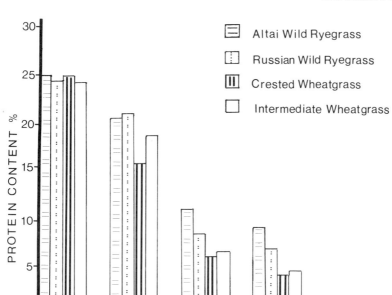

FIGURE 3-16
Percentage of protein content for four grasses at four stages of maturity. (Source: T. Lawrence, 1977, Agriculture Canada Publication 1602)

widespread. It is believed that crossing of the two species and subsequent chromosome doubling occurred in the New England states during the seventeenth century. The new hexaploid plant is much larger than either of its parents and was consequently selected by the settlers because it was more productive. It was first used in New Hampshire and subsequently brought to Maryland by Timothy Hansen, after whom this grass was named.

After being reintroduced to England in 1760, timothy spread rapidly in Europe. In North America, it was carried by the pioneers westward into the Mississippi Valley and northward into eastern Canada. It was taken across the plains and prairies and was extensively used in the mountain valleys of the west. Between 1870 and 1910, timothy was, with oats, the main feed for horses, which were then the primary source of farm power. The increased use of mechanical power from 1910 onward led to a decline in the demand for timothy hay.

Timothy was popular for many reasons, all of which are equally valid today. It is productive, palatable (both in pastures and in hay), and long lived; it grows well with legumes and is easy to establish and harvest. It has a shallow, fibrous root system and requires moderately good rainfall conditions (over 400 mm or 16 in.) for survival. The crown of the plant consists of a group of swollen, bulblike internodes which store carbohydrate material. It produces a mass of light green basal leaves and a long culm with a cylindrical panicle (Figure 3-17), which has a single seed per spikelet (Figure 3-18). At the time when the seed is set, the protein content is low (7% to 8% crude protein), but before maturity, the plant's crude protein content may be as high as 20%. However, when considered on a basis of protein production per unit area of land, timothy yields less than orchardgrass (Table 3-5).

In parts of Wisconsin, timothy will outyield both orchardgrass and smooth bromegrass, but in most other areas the reverse is the case. The disadvantages of this species are associated with its shallow root system. It is not drought resistant and may be uprooted by grazing or by frost heaving. In Canada, *Champ* and *Climax* are widely grown and well-adapted cultivars. *Salvo* and *Zim* are well-known cultivars from Kentucky and Minnesota, respectively.

Wheatgrasses (genus *Agropyron*)

This genus is a part of the *Hordeae* tribe of grasses. The spikelet structure is much the same as in the *Festuceae,* but the many-flowered spikelets are borne on a spike rather than a panicle. The wheatgrasses are found throughout the central and northern parts of the United States, in the intermountain regions of the United States, and on the Canadian prairies. Of the cultivated species, the ones to be considered here are crested wheatgrass [*Agropyron cristatum* L. and *A. desertorum* (Fisch.) Schult.], intermediate wheatgrass [*A. intermedium* (Host) Beauv.], pubescent wheatgrass [*A. trichophorum* (Link) Richt.], slender wheatgrass [*A. trachycaulum* (Link) Malte], streambank wheatgrass (*A. riparium* Scribn. and Smith), and tall wheatgrass [*A. elongatum* (Host) Beauv.]. In general, these species are resistant to drought, easy to establish, and persistent.

Crested wheatgrasses include a diploid (*A. cristatum,* Figure 3-19) and a tetraploid (*A. desertorum,* Figure 3-20) species, which are very much alike. Their $2n$ chromosome numbers are 14 and 28, respectively. Here, the two species will be discussed together, since they are so similar in appearance and agricultural characteristics.

TABLE 3-5
Protein production per unit acre of land for timothy and orchardgrass in Pennsylvania

	Total digestible nutrient		Total digestible protein	
	kg/ha	lb/acre	kg/ha	lb/acre
Timothy	3,272	2,912	756	672
Orchardgrass	4,941	4,398	1,198	1,066

(Source: Mislevy et al., 1977, Agron. J. 69)

FIGURE 3-17
The cylindrical *panicle* found in timothy is so compact that it is frequently mistaken for a spike.

Both species originated in central Asia, eastern Russia, and Siberia. They were first introduced into the United States from Russian Turkestan in 1906 and into Canada from western Siberia in 1911. These species were extensively used in North America in revegetation programs which aimed to combat the dust-bowl conditions generated in the early 1930s.

The crested wheatgrasses are popular because they are readily established and grow well under adverse conditions. The grass grows rapidly in May and June, thus providing grazing when other feeds may be scarce. In pastures, these species respond well to early grazing, which should be heavy for maximum production. Crested wheatgrass may also be used for hay, but the plant's quality and palatability both decline rapidly after heading.

Crested wheatgrass is a bunch species which produces dense roots to a depth of about 2 m (6.5 ft). The plant grows especially well on sandy soils, but will not tolerate alkaline conditions or flooding. It will recover rapidly after periods of drought, during

FIGURE 3-18
Phleum pratense L. Timothy.

FIGURE 3-19
Agropyron cristatum L. Crested wheatgrass.

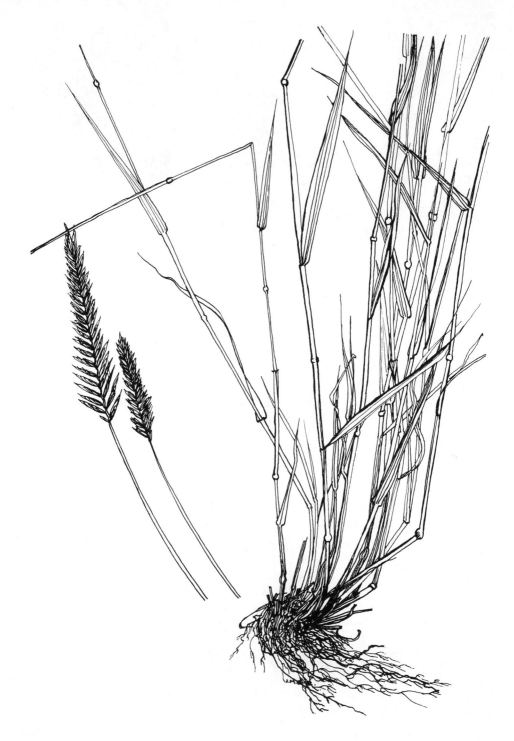

FIGURE 3-20
Agropyron desertorum (Fisch.) Schult. Crested wheatgrass.

TABLE 3-6
Dry matter production from forage species, University of Alberta ranch, Kinsella

	Dry matter yield	
	kg/ha	lb/acre
Intermediate wheatgrass	4,926	4,384
Smooth bromegrass	4,522	4,024
Creeping red fescue	4,206	3,743
Red top	3,888	3,460

(Source: R. H. Gesshe, 1978, M.Sc. thesis)

which it becomes dormant. The species is frequently grown with alfalfa in both pastures and hayland. It is very long lived and winter hardy. Stands 35 to 40 years old are frequently.

The seed production from crested wheatgrass is good, and yields of 1,000 kg/ha (890 lb/acre) are frequently recorded. The inflorescence is a compressed spike. The glumes have short awns. Both species are quite variable, having a range of head types, leaf numbers, and leaf widths. This may be because the crop is highly cross pollinated. *Fairway* and *Parkway*, two diploid cultivars, are widely used in western Canada and the United States. *Summit* and *Nordan* are tetraploid cultivars.

Intermediate wheatgrass was introduced into the United States in 1907, 1923, and again in 1932. Seed from the most recent introduction was brought to Canada in 1935. This species gives high yields on the prairies of western Canada and on the Great Plains of the United States, when used for either hay or pasture. At the University of Alberta ranch, intermediate wheatgrass outyields smooth bromegrass (Table 3-6).

A performance of this type can be expected from intermediate wheatgrass only where the rainfall is 350 mm (14 in.) or over, since the species does not yield well under very dry conditions. Although the plant is deep rooted and forms a rhizomatous turf, it is not as aggressively creeping as smooth bromegrass (Figure 3-21). Intermediate wheatgrass is not as long lived or as winter hardy as crested wheatgrass. Trials at the University of Alberta ranch have also shown it to be less palatable than many species commonly grown in somewhat dry areas (Table 3-7).

The seeds of intermediate wheatgrass are quite large, making this species easy to establish. The cultivar *Chief* was released in 1961 in Saskatoon. *Oahe*, another cultivar commonly used, was bred in South Dakota.

Pubescent wheatgrass (Figure 3-22) was introduced into North America from the USSR at about the same time as intermediate wheatgrass, which it closely resembles in appearance, area of origin, and agricultural characteristics. The two species are readily distinguished from one another by the short stiff hairs covering the head and seed of pubescent wheatgrass (Figure 3-23). It is only in very recent years that the agricultural value of pubescent wheatgrass has been appreciated. Pubescent wheatgrass

FIGURE 3-21
Agropyron intermedium (Host) Beauv. Intermediate wheatgrass.

FIGURE 3-22
Agropyron trichophorum Link. Pubescent wheatgrass.

TABLE 3-7
Fall utilization of forage species for animals given free choice, University of Alberta ranch, Kinsella

	Dry matter utilization	
	kg/ha	lb/acre
Red top	3,495	3,111
Smooth bromegrass	2,805	2,497
Russian wildrye	1,000	890
Intermediate wheatgrass	942	838

(Source: R. H. Gesshe, 1978, M.Sc. thesis)

is considered to grow better than intermediate wheatgrass where soils are alkaline or of low fertility and where rainfall is low. The seed of pubescent wheatgrass is somewhat smaller than that of intermediate wheatgrass. *Greenleaf* is the only cultivar licensed for sale in Canada. It has good seedling vigor, as well as high seed and forage yield. A well-known cultivar, widely used in the United States, is *Mandan 759,* which was developed in North Dakota.

FIGURE 3-23
The spike of pubescent wheatgrass.

Slender wheatgrass orginates in southwestern Canada and the northern parts of the United States. There are a number of improved strains which have been selected from native stocks and are now licensed as cultivars. The plant has a bunch habit, with dense fibrous roots about 40 cm (14 in.) deep. It is very short lived and frequently dies out of a pasture after 3 or 4 years. It grows best on moist but well-drained soils and is tolerant of some alkalinity and drought. It establishes well and develops a good ground cover quickly, but it will not withstand heavy grazing. Slender wheatgrass grows well in a mixture with alfalfa but will not compete with crested wheatgrass or quackgrass. It has a good nutrient composition and is reasonably palatable. The plant has a short ligule and a small auricle. The inflorescence is an erect spike with glumes and lemmas which are awnless or nearly so (Figure 3-24). Unlike most forage grasses, it is self-pollinated.

The two most widely used cultivars are *Revenue* and *Primar*, bred at an Agriculture Canada Research Station and by the Agronomy Division of Washington Agricultural Experiment Station, respectively. *Revenue* has a higher leaf-to-stem ratio and better herbage and seed yield than *Primar*. It also is superior to *Primar* in terms of dry matter digestibility.

Streambank wheatgrass is a species native to North America. It is a long-lived, rhizomatous, sod-forming grass which establishes a full ground cover quicker than any other dryland species (Figure 3-25). For this reason, and as the name suggests, this species is frequently used for erosion control and for stabilizing canal banks and waterways. It is also used on roadsides, airstrips, and in yards. For grazing, the yield is low, but the species is alkali tolerant and drought resistant. *Sodar* is the name of the cultivar most commonly used. It was bred jointly by workers in Idaho and Washington.

Tall wheatgrass is a native of the salt flats and seashores of southern Russia. There are records of two introductions into North America. In 1929, tall wheatgrass was brought to the University of Saskatchewan in Canada, and in 1932 it was introduced into the central part of the United States. In both countries, the species initially attracted little attention, since it is slow to establish and not very drought resistant. However, the discovery that it survives well on saline soils and competes with useless and weedy species that would otherwise dominate these areas has ensured tall wheatgrass an important, if somewhat limited, place in western agriculture. The species is now grown in the central and western part of southern Canada and the northern United States.

Tall wheatgrass is a coarse-leafed, late-maturing bunch grass with short ligules and a small auricle (Figure 3-26). The plant grows slowly when young. The inflorescence is a large spike 12 to 25 cm (4 to 9 in.) long. Frequently, the lower spikelets recurve and do not overlap (Figure 3-27). The glume is square at the top and the seed is large.

In spite of its coarse leaves, this species is palatable and makes very good hay if cut shortly after heading. It also provides good pasture, growing well with alsike clover and producing fair yields. The cultivar used in Canada is *Orbit*. This cultivar is more winter hardy than *Alkar*, which was released in Washington and is widely used in reclamation work in the northern United States.

FIGURE 3-24
Agropyron trachycaulum (Link) Malte. Slender wheatgrass.

FIGURE 3-25
Agropyron riparium Scribn. and Smith. Streambank wheatgrass.

FIGURE 3-26
Agropyron elongatum (Host) Beauv. Tall wheatgrass.

FIGURE 3-27
Tall wheatgrass showing recurved spikelets.

WARM-SEASON PERENNIAL GRASSES

The warm-season grasses, which originate in Africa, southeast Asia, and South America, have either the C_3 (Calvin) or C_4 (Hatch and Slack) photosynthetic pathways. They require warm temperatures for optimum growth and in general are not winter hardy, but can tolerate high temperatures and periods of drought. When grown for hay, they produce several crops per year, giving higher yields of poorer-quality forage than do the cool-season grasses. They require short, warm days for floral induction and produce several seed crops per year.

In the humid southeastern regions of the United States, the warm-season perennial forage grasses provide high yields between early May and late August. Examples are bahiagrass, bermudagrass, dallisgrass, carpetgrass, johnsongrass, napiergrass, and pangolagrass. Such species are used in conjunction with cool-season perennial grasses (Kentucky bluegrass, tall fescue, and orchardgrass). Cool-season grasses produce forage from early March to late May and for the months of September and October. For the remaining months of the year (January, February, November, and December), ryegrass or small-grain crops are used for forage production.

Temperature and moisture interact to influence forage production in tropical and semitropical regions. For the warm-season grasses, the optimum growth temperature is 30° to 35°C (86° to 98°F), with growth stopping at about 15°C (59°F). This contrasts with the cool-season species, whose optimum growth temperature is 20°C (68°F), and which cease growth at 5° to 10°C (41° to 50°F). Moisture, in combination with these temperatures, then determines the range of adaptation for a warm-season species.

Some cool-season grasses, such as tall fescue, will grow on heavy soils in areas where summer temperatures are high and, under these circumstances, irrigation results in only small yield increases. Where rainfall is low and soil is light, as in eastern Texas, bermudagrass and dallisgrass are the most widely grown species. In the dry areas of Arizona, New Mexico, and western Texas, big bluestem, little bluestem, Indiangrass, and switchgrass are widely used native species. Blue panicgrass and buffelgrass are cultivated species which are grown in this area.

On irrigated pastures in the desert valleys of southern California, bermudagrass and sudangrass are commonly grown. Adapted cultivars of cool-season grasses (orchardgrass, tall fescue, perennial ryegrass, timothy, and meadow foxtail) are used west of the Cascade Mountains in the United States in the subhumid areas having a continental climate. For such areas, production is much improved by applications of a nitrogenous fertilizer.

During the last 25 years, considerable emphasis has been placed on the use of cultivated grass and legume pastures in the humid subtropics of the southeastern United States. Especially in years of lower than normal rainfall, yields of row crops such as cotton, corn, tobacco, and peanuts are enhanced where these crops follow semipermanent pasture. Bermudagrass is the most effective grass in increasing soil organic matter when the pasture is plowed, while the shallow-rooted carpetgrass *(Axonopus affinis)* is the least effective. In the semiarid southern Great Plains, temporary pastures of sudangrass are used in rotations on unirrigated land. On irrigated rice lands, *Paspalum dilatatum* is grown for 3 or 4 years, following continuous rice cropping for 3 to 10 years.

TABLE 3-8
Quality characteristics of cool- and warm-season grasses in North Carolina

	Acid-detergent fiber %	Lignin %	Digestibility %
Timothy	31.2	5.8	63.9
Tall fescue	21.6	2.3	78.1
Sorghum	32.3	5.9	64.6
Bermudagrass	36.8	7.7	57.7

(Source: Burns and Smith, 1980, Agron. J. 72. By permission of The American Society of Agronomy, Inc.)

TABLE 3-9
Average crude protein content and yield of subtropical grasses harvested at 3-month intervals, Florida

	Crude Protein %	Dry Matter T/ha	tons/acre
Cynodon spp.	14.3	4.8	2.6
Paspalum spp.	14.0	2.3	1.1
Digitaria spp.	15.3	4.0	1.8

(Source: Misley and Everett, 1981, Agron. J. 73. By permission of The American Society of Agronomy, Inc.)

The warm-season grasses, especially when grown in the tropics, are poor in quality when compared with well-managed pastures in temperate areas (Table 3-8). Dry-matter yields are often high, but the material is very fibrous, has a low digestibility, and has a poor protein content (Table 3-9). Under these circumstances, legume species make a very important contribution to forage quality. For example, in the absence of a legume, yields of dry matter for native warm-season grass pastures range from 400 to 4,000 kg/ha (356 to 3560 lb/acre), with a crude protein content of between 3.0% and 3.7%. When only material of this type is available, high and intensive production from dairy cows or young beef animals is impossible.

Many warm-season perennial grasses produce little or no seed and are propagated vegetatively. Bermudagrass, napiergrass, and pangolagrass are examples of species established in this way.

Bahiagrass (*Paspalum notatum* Fliigge)

Many members of the genus *Paspalum* originate in the southeastern United States. Bahiagrass is an exception, being native to South America, where it is widely used in Argentina, Brazil, Mexico, Paraguay, and Uruguay. The species was introduced into Florida at the beginning of this century, and by 1915, tests indicated that it was a productive, palatable, and nutritious perennial which could spread from natural seeding. It has a number of Spanish names, of which *Pasto horqueta, Gama dulce,* and *Gengibrillo* are the most commonly used. Bahiagrass is used extensively in yards and playing fields and around public buildings and homes in east Africa and in India. In the United States, it is grown in the southeast, mainly in the coastal areas, and to a lesser extent inland to east Texas and north to the Arkansas. The main agricultural value of bahiagrass is for permanent pasture, where it persists well, withstanding both grazing and trampling. Bahiagrass spreads rapidly by short rhizomes, forming a dense sod, with extensive fibrous roots to depths of 2.5 m (8.25 ft) (Figure 3-28). It readily becomes sod bound, and its aggressive root system suppresses legumes when grown in a mixture. Regular applications of nitrogen fertilizer are required to maintain productivity at a high level. Bahiagrass will survive, however, under low soil fertility conditions, providing soils are not heavy clays or inclined to be wet. It starts growth early in the spring,

FIGURE 3-28
Paspalum notatum Flügge. Bahiagrass.

remaining green late into the fall. It is very sensitive to cold, but is free from serious disease or insect problems.

Bahiagrass is normally established from seed early in the spring. It produces seeds which have a hard seed coat and waxy glumes, and which germinate slowly unless scarified by mechanical means or by sulfuric acid. The seed will remain viable for many years in the soil and also after passing through the digestive system of animals. This, combined with the aggressive nature of bahiagrass, enables the plant to take over a pasture again after it has been plowed down and the area seeded to another species. Culms are about 45 cm (1.33 ft) long, and racemes are curved with spikelets in two rows on one side of a primary branch (Figure 3-28).

In general, the yields of both forage and beef from bahiagrass pastures are lower than for those planted with bermudagrass hybrids. Bahiagrass can withstand close grazing and, under those circumstances, gives good-quality herbage. Seed yields are usually good; the spring burning of trash from the previous season before growth starts has been shown to increase seed yields. The application of a nitrogenous fertilizer 1 to 2 months before the seed is harvested also increases seed production.

Bahiagrass may be divided into a number of types on the basis of its origin and leaf size. The short, narrow-leafed types, being tough and low yielding, are unsuitable for pasture. Such cultivars are used in yards and on roadways, airstrips, and highway shoulders. Cultivars with long, broad leaves are used for agricultural purposes. *Pensacola* is a cultivar extensively used in the southern United States. It has some winter hardiness, good seed germination, and responds well to nitrogenous fertilizers, but is now being replaced with new cultivars from the United States Department of Agriculture in Georgia.

Bermudagrass (*Cynodon dactylon* L. Pers.)

Bermudagrass is also known as *couch, stargrass, gros chiendent* (French), and *pasto bermuda* (Spanish). The species is now grown throughout the tropics and subtropics. There is some doubt as to its origin. It is described by various authors as originating in India, Africa, or the Mediterranean. Today it is of considerable importance in the southern United States, where it is grown from coast to coast and is called the "king of pasture plants." Its areas of growth are limited only by the fact that temperatures of about $-2°C$ (28.4°F) will induce dormancy. On a world scale, it is grown on irrigated land in tropical or subtropical areas or where there is a rainfall of between 650 and 1,750 mm (26 and 70 in.) per year.

Bermudagrass is an excellent perennial pasture plant. It establishes from stolons, rhizomes, and seed. It does well on deep fertile soils and can grow on both heavy clays and sandy soils. It withstands some flooding, but does not grow well on waterlogged soils. It is acid, alkali, and somewhat salt tolerant. In a pasture, bermudagrass responds very well to fertilizer applications and withstands close continuous grazing and trampling. In the earlier stages of growth, both palatability and protein content (12% crude protein) are good and compare favorably with johnsongrass and dallisgrass. The percentage of digestibility of bermudagrass is 65% or higher. The taller strains cure well and produce an excellent hay. This species is also valuable for soil-conservation purposes. When established from seed, it is usual to plant between 9 and 11 kg/ha (8 to 10 lb/acre). However, because the most productive cultivars are sterile hybrids, bermudagrass is usually established from *sprigs*, or *stolon cuttings*, which are best obtained fresh from the farm's own nursery area. At planting, part of the sprig is left exposed and the soil is well packed around the remainder of the vegetative material. A nitrogenous fertilizer should be applied when the new stolons reach a length of 12 to 13 cm (4 to 5 in.).

The high cost of establishment is justified, since bermudagrass stands are very long lived and, where management practices are sound, yields are high. When mature, the nutritive value of bermudagrass is low and the herbage is unpalatable. Consequently, it is important that stocking rates be high and that the plants be maintained in the juvenile stage (see Chapter 17).

The vigorous nature of the bermudagrass plant makes it difficult to maintain legumes in a sward. However, both winter and summer legumes may be seeded or reseeded into established bermudagrass pastures. The most commonly used winter legumes are vetches or clovers (crimson, red, arrowleaf, or white clover); lespedeza or alfalfa are commonly used as summer legumes. Where legumes are grown in this way, grazing management should be directed toward favoring legume growth. Grazing should be close for low-growing legumes like white clover, while taller-growing species like alfalfa should have periods of rest.

Depending on the nature of the landmass, bermudagrass may be used as a summer perennial within the northern and southern limits of the tropics. For some considerable distance north and south of these lines, it is possible to use bermudagrass as a summer perennial and overseed it with a winter annual grass (e.g., Italian ryegrass), thus obtaining year-round production. This practice is especially successful where forages are being grown under irrigation. In the dry areas of northern Mexico and in the southern United States, as irrigated bermudagrass becomes dormant with the onset of cooler weather in the fall, it is overseeded with a small-grain cereal crop (oats or barley), Italian ryegrass, or both. Seeding rates are 90 kg/ha (80 lb/acre) of the small-grain crop and 17 kg/ha (15 lb/acre) of the annual ryegrass. Excess bermudagrass is removed before planting by grazing or haying. Fescues have also been used for overseeding bermudagrass. Since fescue seedlings are small, it may be necessary to use a leaf contact herbicide, just before seeding, to kill the bermudagrass top growth and reduce competition. The practice of overseeding normally reduces the yield of bermudagrass, but this depends on the overseeded species and grazing management. However, total annual production is substantially increased, and since the winter is a time when feed is less readily available, this gain will frequently outweigh the loss.

The inflorescence of bermudagrass is a multiple spike between 15 and 30 cm (5 and 10 in.) long (Figure 3-29). The leaves are gray to bluish-green, short, and flat. The seeds are very small; there are between 3 million to over 4 million seeds per kilogram (1.3 to 2 million seeds per pound). Germination is frequently poor.

Many of the bermudagrass cultivars now in common use were bred by Glenn W. Burton at the Georgia Coastal Plain Experiment Station. *Coastal, Midland,* and *Coastcross-1* are all hybrids. Coastal is a very productive variety that responds well to fertilizer applications. It is resistant to diseases, including root-knot nematodes. Midland is winter hardy and is grown at the northern limits of the bermudagrass area of

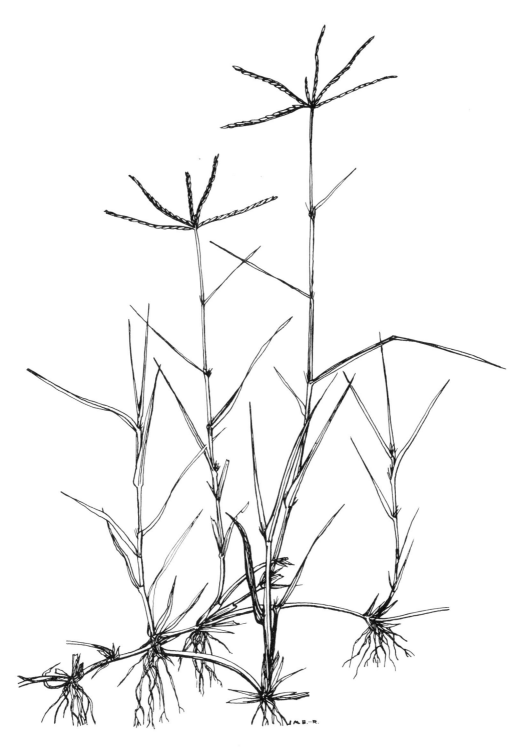

FIGURE 3-29
Cynodon dactylon (L.) Pers. Bermudagrass.

adapation (Kansas, Maryland, and Missouri). Coastcross-1 (also called *Cross-1*) is a sterile hybrid which gives very high production, but is more sensitive to low temperature than Coastal. It has rapidly growing stolons and no rhizomes. The forage production is about the same as Coastal, but the material is much more digestible, thus giving higher animal-weight gains. *Callil* and *Tifton 44*, both of which are highly digestible, are new cultivars.

Carpetgrass (*Axonopus affinis* Poir)

Carpetgrass is used in the southern United States in Texas, Virginia, Arkansas, Alabama, Mississippi, and Louisiana. It was introduced to this region from the West Indies and Central America. This low-growing species spreads by stolons to form a dense sod with abundant roots. The plant has many basal leaves. As compared with other perennial summer grasses, its nutritional value and yield is low. It will not thrive in swamps, but does well where moisture is abundant and soils are sandy or well drained. In the lower south of the United States, it is frequently found along creeks and rivers where water tables are high. It is adapted to soils of low fertility.

Trials in Georgia indicate that carpetgrass pastures give about one-third the beef production of common bermudagrass. Large areas consisting of carpetgrass, native bluestems, and broomsedge in the Gulf Coast and in the southern states on the Atlantic Coast have a very low carrying capacity (5 to 15 ha or 12.5 to 37.5 acres per animal unit). Carpetgrass is used for lawns, golf courses, roadsides, waterways, ditches, and firebreaks. It is not used very frequently in pastures because it is difficult to maintain legumes in a carpetgrass sward.

Dallisgrass (*Paspalum dilatatum* Poir)

Dallisgrass originates in southern Brazil, Argentina, and Uruguay, where it is known by its Spanish names *pasto miel* or *gramma de agua*. It was first introduced to the humid south of the United States in 1842. In 1875, it was reintroduced by A. T. Dallis, who gave the plant its present name. The species is also grown in Australia, New Zealand, Hawaii, and South Africa.

Dallisgrass has a bunch-type growth habit (Figure 3-30) with a large number of basal leaves. It has a deep root system, with some rhizomes. It is drought tolerant and will tolerate wet soils better than bahiagrass. It is an excellent permanent pasture grass which gives spring growth more palatable than bermudagrass. Not only is it the first species to produce spring growth in the lower southern United States, but it also gives high late-season yields and has a longer growing season than any other warm-season grass. It is very persistent and withstands both trampling and close grazing well. Its buched growth habit enables it to combine with legumes and other grasses; it is widely grown with clovers and lespedeza. Dallisgrass and white clover mixtures are usually found in low wet areas. Since dallisgrass is rather slow to establish, a well-prepared and weed-free seedbed is essential. The cultivar *Prostrate*, developed in Georgia, is well adapted to light and poorly drained soil. The seed quality is very low, since it is an obligate apomict and very susceptible to ergot fungus.

FIGURE 3-30
Paspalum dilatatum Poir. Dallisgrass.

Johnsongrass (*Sorghum halepense* L. Pers.)

Johnsongrass is found extensively as a native species in India, Asia, and the Mediterranean region. In the United States, it is an important perennial in the Southeast. It was introduced into that region from Turkey in 1830. It derives its name from Colonel W. Johnston, who introduced it into Alabama in 1840.

It is regarded as a serious weed by row-crop farmers, but stockmen recognize the importance of its high yielding ability, vigor, good nutritive value, and persistence. It is the only perennial member of the genus *Sorghum* and has large tenacious rhizomes and seeds that will pass through livestock. Both these characters make johnsongrass difficult to eradicate (Figure 3-31). Johnsongrass is adapted to well-drained, heavy, black-clay soils, but will grow well on most soils. It makes good hay if cut early and responds to fertilizer applications with substantial yield increases and high crude protein content. Johnsongrass will not withstand continuous grazing at high stocking rates but it has proved to be persistent under rotational grazing, with 1 week of use and 3 or 4 weeks of rest.

Pangolagrass or Digitgrass (*Digitaria decumbens* Stent.)

Pangolagrass, or digitgrass, is a native of South Africa, where it grows well in areas with 625 to 750 mm (25 to 30 in.) of rainfall. It is used as a pasture grass throughout the Caribbean and is a component of many high-yielding pastures in that region. It was introduced to the United States in 1902 and is now grown in southern Florida and on the Gulf Coast of Mexico. The name "digitgrass" is also used for *Digitaria pentzii* Stent., slenderstem digitgrass, and other species in the genus *Digitaria*.

Pangolagrass is a creeping perennial with semidecumbent runners which root at the nodes. The seedstalks are many and the number of viable seeds are few. It is not winter hardy. Pangolagrass is used mainly in pastures and gives better animal-weight gains than either bermudagrass or bahiagrass (Table 3-10). The herbage has a high protein content, and the plant yields well where rotationally grazed at intervals of about 1 week. If the plant is allowed to grow 30 to 45 cm (12 to 18 in.) high before grazing, it is able to persist with heavy use. Digitgrass may be used to make excellent hay or silage. This species is established vegetatively from runners.

TABLE 3-10
Average daily gains and beef production from warm-season grasses in Florida

	Average daily gain		Gains	
	kg/day	lb/day	kg/ha	lb/acre
Digitgrass	0.28	0.61	461	410
Bahiagrass	0.26	0.57	407	362

(Source: Adjei et al., 1980, Agron. J. 72. By permission of The American Society of Agronomy, Inc.)

FIGURE 3-31
Sorghum halepense L. Pers. Johnsongrass.

FIGURE 3-32
Chloris gayana Kunth. Rhodesgrass.

Rhodesgrass (*Chloris gayana* Kunth)

Rhodesgrass is a native species of south and east Africa which was introduced into the United States in 1902. It is limited in its areas of use, since it is killed by heavy frost. It will, however, grow vigorously and produce heavy yields in Florida and south Texas. This species is adapted to a wide range of soil types, growing well on alkaline soils. It is a palatable pasture grass that withstands grazing and trampling well. Rhodesgrass will establish easily from seed, and since it quickly forms a good soil cover, it may be used for erosion control. It spreads by both seed and stolons, but is easily controlled in row crops (Figure 3-32).

Saint Augustinegrass (*Stenotaphrum secundatum* Walt.)

Saint Augustinegrass is also known as buffalograss in Australia and crabgrass in the Caribbean. This species is well adapted to coastal areas in the humid tropics and subtropics. It is found in coastal districts of North, Central, and South America, the West Indies, West and South Africa, Asia, Australia, and the Pacific Islands.

The plant is a broad-leaved, creeping, coarse, glabrous perennial. The stolons are long, with branches which are flat, short, and leafy, thereby forming a dense sod. It is essentially a pasture plant. Yields of dry matter are high, with a crude protein content of about 15% when the plant is kept in the juvenile stage. The species is palatable, persistent, and able to withstand trampling and some shading. As the herbage matures, it rapidly becomes unpalatable. Pasture productivity can be maintained by grazing at 12- to 14-day intervals to a height of 5 to 7.5 cm (2 to 3 in.). Pastures are slow to recover if grazed too closely. Establishment of St. Augustinegrass is from stem cuttings. It takes from 5 to 6 months to achieve full ground cover. This species is frequently grown on muck soils in the Everglades in Florida.

FURTHER READING

Ahlgren, G. H., C. Eby, 1942, Better pasture and hay crops, N.J. Agr. Exp. Sta. Circ. 448.

Best, K. F., J. Looman, and J. B. Campbell, 1977, Prairie Grasses, Agriculture Canada Publication 1413, Ottawa.

Campbell, J. B., K. F. Best, and A. C. Budd, 1966, 99 Range Forage Plants of the Canadian Prairies, Agriculture Canada Publication 964, Ottawa.

Elliot, C. R., and H. Baenzigor, 1973, Creeping Red Fescue, Agriculture Canada Publication 1122, Ottawa.

Elliott, C. R., and J. L. Bolton, 1970, Licensed Varieties of Cultivated Grasses and Legumes, Agriculture Canada Publication 1405, Ottawa.

Goplan, B. P., S. G. Bonin, W. E. P. Davis, and R. M. MacVicar, 1963, Reed Canarygrass, Agriculture Canada Publication 1963, Ottawa.

Gould, F. W., 1968, Grass Systematics, McGraw-Hill Book Company, New York.

Hitchcock, A. S., 1951, Manual of grasses of the United States, U.S.D.A. Misc. Publication 200, rev.

Hoover, M. M., M. A. Hein, W. A. Dayton, and C. O. Erlanson, 1948, The main grasses for farm and home. *In* Grass, U.S.D.A. Yearbook Agr., pp. 639–700.

Knowles, R. P., and E. Berglass, 1971, Crested Wheatgrass, Agriculture Canada Publication 1295, Ottawa.

Lawrence, T., 1977, Altai Wild Ryegrass, Agriculture Canada Publication 1977, Ottawa.

Lawrence, T., and D. H. Heinrich, 1977, Growing Russian Wild Ryegrass, Agriculture Canada Publication 1607, Ottawa.

Leithead, H. L., L. L. Yarlett, and T. M. Shiflet, 1971, 100 Native forage grasses in 11 southern states, U.S.D.A. Handbook 389.

Lodge, R. W., S. Smoliak, and A. Johnson, 1972, Managing Crested Wheatgrass Pasture, Agriculture Canada Publication 1473, Ottawa.

Newell, L. K., 1955, Wheatgrass in the west, Crops and Soils 8:7–9.

Piper, C. V., 1942, Forage Plants and Their Culture, rev. ed., pp. 202–214, Macmillan, New York.

Vose, P. B., 1959, The agronomic potentiality and problems of the canarygrasses, Herb Abstr. 29:77–83.

chapter 4

The Perennial Legumes

Legumes have been in agricultural use for at least 6,000 years and are a group of plants of the utmost value to farmers throughout the world. Lake dwellers in Switzerland used legumes in 4,000 B.C., and Chinese literature provides information on the cultivation of soybeans in 2,500 B.C. Legumes were also used by the early Egyptians and by the Romans. Today, legumes are found wherever agricultural settlements have developed. As might be expected, most species grow best in specific regions, and, within the species, strains have developed which are adapted to a particular range of climatic and soil types.

The seed sizes for some common legumes are given in Table 6-4. Following the same format as that of the grasses, we will consider here first the legumes used in temperate areas and then those used in the tropics and semitropics.

SPECIES USED IN TEMPERATE AREAS

Alfalfa (*Medicago* spp.)

Although alfalfa has been placed under this heading because of its wide usage in temperate areas, it is, in fact, grown in both temperate and subtropical locations. On a global basis, it is not only the most widely used forage, but also the oldest. The center of origin of this herbaceous perennial is in Iran, Transcaucasia, and Asia Minor. Alfalfa has evolved in response to the cold winters and hot, dry summers of these regions.

As well as common alfalfa (*Medicago sativa* L., Figure 4-1), there is an exceptionally cold-tolerant species (*M. falcata* L.), which has been used under the growing conditions prevailing in the far north of Siberia. Both alfalfa species were introduced into Germany and northern France about 450 years ago. It is believed that there hybridization took place between the blue to purple-flowered *M. sativa* and the yellow-flowered *M. falcata*. The resultant hybrids are called *M. media* and have to varying degrees the characteristics of their two parents. The flowers of the hybrid species are frequently variegated, the pods are twisted to form one-half to five spirals, and the seeds are kidney shaped. The new species has great genetic variability, which, combined with various voyages of discovery initiated in Europe over the sixteenth and seventeenth centuries, has helped to establish alfalfa throughout the world. Plants of both *M. sativa* and *M. media* are adapted to growing conditions in North and South America, Australia, New Zealand, South Africa, the cooler regions of Asia, the Mediterranean area, and northern Europe. There are a few cultivars of *M. falcata* which are used infrequently (e.g., a cultivar called Anik, grown in northern Alberta). The most commonly used names (which are applied to all three species) are *lucerne* (in Europe, South Africa, Australia, Africa, and New Zealand), *alfalfa* (in North and South America and Asia), and, sometimes, the *medics*.

Alfalfa was introduced into South America by the Portuguese and Spaniards in the sixteenth century. It was moved from Peru and Mexico to Texas, Arizona, New Mexico, and California by missionaries, and was well established in these areas by 1870. From this beginning, it has come to be known as "the queen of forages" and now occupies 13 million ha (33 million acres) in North America and 33 million ha (83 million acres) on a world scale. Especially in terms of protein production, it has outstanding feeding value, surpassing all other forage crops.

Within the *M. sativa-falcata* genetic complex, there is so much physiological and morphological variability that it is meaningless to discuss the regions of adaption of the group as a whole. With the exception of those tropical areas where temperatures are consistently high, alfalfa has cultivars, or groups of cultivars, which grow in all agricultural situations. There are a number of reasons for this great range of diversity. First, there are both diploid ($2n = 16$) and tetraploid forms ($2n = 32$), which will cross to give fertile hybrids when the diploid is raised to the tetraploid level. Second, individual plants are almost entirely cross-pollinated, so the level of heterozygosity in a population is high. Third, this vast background of genetic variation has been exposed to natural selection on a global scale for between 3,000 and 4,000 years. Consequently, alfalfa cultivars are normally divided into four main groups:

1. The *common* group contains only *M. sativa* types, which are all purple flowered. This group includes Hungarian, French, Italian, and Spanish alfalfa cultivars, as well as strains grown in Argentina, South Africa, New Zealand, and the United States (i.e., *Buffalo* and *Cody*).

2. The *Turkistan* group contains those *M. sativa* types which originate in

FIGURE 4-1
Medicago sativa L. Common alfalfa. Only the ends of the leaflets are "toothed"; cf. sweetclover.

central Asia. They are cold-resistant, give low seed yields, recover rather slowly from cutting, and start winter dormancy early. Some of these have been introduced into North America and successfully selected for winter hardiness.

3. The *variegated* group consists of cultivars of *M. media*. The flower colors include white, cream, yellow, blue, blue-green, and purple. This group contains a large number of well-known cultivars, many of which originated in France or Germany. *Province* has rather poor winter hardiness; it is from southeastern France. There are many *Flemish* cultivars grown in northern France, Scandinavia, and Great Britain. *Grimm* is a well-known, high-yielding, winter-hardy, and drought-resistant cultivar from which a number of newer cultivars have been selected. *Cossack, Ladak, Ranger, Rambler, Atlantic,* and *Narragansett,* all members of the variegated group, are extensively grown in North America.

4. The *nonhardy* group has cultivars which have a very erect growth habit, recover well after cutting, and have a long growing season with or without a short period of dormancy. They are used in the semitropics, frequently under irrigation. They are adapted to short days and are susceptible to low temperatures. The best-known strains within this group are *Peruvian* (smooth and hairy), *Indian* (grown in Poona and Bombay states), *Egyptian,* and, from Argentina, *Rio Negro, Saladina,* and *Faculdad San Martin.*

Alfalfa may be used for grazing (see Chapter 17) or for hay (see Chapter 11), or it may be artificially dried or dehydrated. Following defoliation, it starts regrowth relatively quickly, but thereafter builds up its carbohydrate reserves rather slowly. Consequently, if it is defoliated frequently and at short intervals, carbohydrate reserves are depleted and the plant dies. After grazing, pastures containing alfalfa should be rested. Also, to allow carbohydrate reserve levels to recover, alfalfa should not be cut prior to the onset of winter. This is a time when the plant's activities are slowed by the onset of dormancy and when it must be permitted to build up its reserve of carbohydrate material for respiration during the winter. Cutting 3 to 4 weeks before the first killing frost may result in the plant failing to survive the winter. These considerations are important in relation to grazing management and when harvesting alfalfa for use in commercial dehydrating operations.

Alfalfa may be made into a silage high in protein. Good-quality alfalfa silage is, however, much more difficult to make than is corn silage, since protein is one of the first substances to be broken down if the silage-making process is not efficiently managed. This problem may be overcome by mixing corn or sorghum with alfalfa when the silo is filled. The moisture content for silage should be between 60% and 70%. If an airtight silo is available, alfalfa haylage (40% to 60% moisture) of very high quality may be made.

In spite of its wide range of adaption, alfalfa does not grow well where soils are at all acidic. It seems that it is the bacteria (genus *Rhizobium*) present in the root nodules, rather than the plant itself, which are sensitive to pH values below 6.5. Because the alfalfa root system is very large, with a long taproot which will penetrate the soil some 3 to 6 m (10 to 20 ft), the plant is able to draw moisture and nutrients from considerable depths. The part of the taproot just below the crown forms a thick, fleshy storage region for carbohydrate material.

Alfalfa is an upright plant, growing 1 to 3 m (10 to 20 ft) in height. It grows best with bunch grass species (e.g., orchardgrass) and may be crowded out of forage mixtures which include aggressive rhyzomatous grasses. There are some alfalfa cultivars with "creeping" stem systems. These are rather prostrate plants which spread slightly by underground stems. The cultivars *Rambler, Roamer,* and *Drylander* show these characteristics and are winter hardy. There are over 100 alfalfa cultivars in common use, each of which having their areas of adaptation. To name but a few, *Vangard* is adapted to live in southeastern United States, *Titan* is hardy, and *Granda* has disease and insect resistance. The growth rates of alfalfa grown in New York (Table 4-1) show that this plant cannot withstand flooding for more than a few days. Winter survival in alfalfa is not only a matter of cold tolerance. Diseases such as crown rots, which are of bacterial and fungal origin, make a substantial contribution to winterkill. In fact, few plant deaths during the winter are due to one factor only, and it is frequently difficult to determine what is the main cause of a decline in plant population.

Birdsfoot Trefoil (*Lotus corniculatus*)

Little is known of the origin and early use of birdsfoot trefoil. It is widely distributed in Europe and was first described there in 1597. How it reached North America is equally uncertain. There is some evidence to suggest that it was introduced by chance about 130 years ago. The use of the plant as a cultivated species was first recorded in both Europe and North America about 1900.

Birdsfoot trefoil stems are much finer than those of alfalfa and tend to be prostrate (Figure 4-2). It is the only forage legume with a leaf consisting of five leaflets (see *f* in Figure 10-1). The seedpods, which radiate from a common axis, look like a bird's

TABLE 4-1
Growth rate of flooded and unflooded alfalfa plants in New York

Days flooded	Relative growth rate
1[a]	0.88
2	0.77
3	0.65
4	0.52
5	0.43
6	0.20
8	0.00

[a](1.0 = yield not reduced by flooding)

(Source: Tompson and Fick, 1981, Agron. J. 73. By permission of The American Society of Agronomy, Inc.)

84 / The Perennial Legumes

FIGURE 4-2
A dense cover of birdsfoot trefoil viewed from above.

foot (Figure 4-3), hence the species' name. Its root system is considerably smaller than that of alfalfa. It has, however, very good drought resistance (i.e., better than ladino clover), and some cultivars are winter hardy. Compared with alfalfa, it is more acid tolerant and will grow on poorly drained soils. It may be used for hay, but is normally grown as a pasture plant. Because seed of birdsfoot trefoil germinates slowly, and seedling growth is slow as well, it frequently takes 2 years to establish a new pasture. During this time, weed growth provides heavy competition for birdsfoot trefoil. Once established, however, the pasture will be long lived, due in part to the way in which the mature pods break open and shed large numbers of seed, thus facilitating reseeding.

Birdsfoot trefoil grows and yields well during the summer months. Maintenance of the species in a pasture does not require the same skill in pasture management as does alfalfa, for it will regrow from its own seed. It is, however, 10% to 15% lower yielding than alfalfa. Birdsfoot trefoil is palatable and has a high nutritive value. Its most important quality characteristic, however, is that it has never been known to cause bloat (see Chapter 10). *Viking* is a most widely used cultivar of this species as well as *Empire, Leo,* and *Maitland,* bred in New York, Quebec, Guelph, and Ontario, respectively.

True Clovers (genus *Trifolium*)

The members of this genus are all herbaceous, have trifoliate leaves, and have many small flowers held on a short, crowded floral head. The standard is only slightly larger

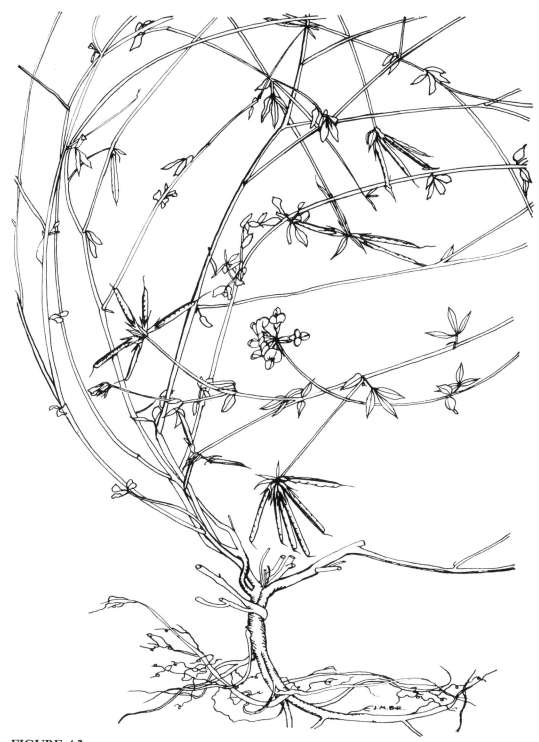

FIGURE 4-3
Lotus corniculatus L. Birdsfoot trefoil.

than the other petals. The true clovers may be annuals, biennials, or perennials, and are found in humid temperate areas of the northern hemisphere.

Alsike Clover (*Trifolium hybridum* L.). This clover derives its name from the parish of Alsike in central Sweden. It was here, in 1750, that its agricultural use was first recorded. Apart from this, its origin is uncertain. It was grown in England and Scotland in 1832, and was introduced to the United States and eastern Canada in 1839.

Although alsike clover is a true species and not a hybrid, as its Latin name suggests, it appears to be intermediate between white and red clover. It is a short-lived perennial, but is frequently used as a biennial. It has pink flowers on branches which arise from each leaf axil (Figures 4-4 and 4-5). The flowers curve downward after pollination and turn brown as the seed matures.

Alsike clover grows on many different soil types. It tolerates soil which is too acid for red clover and too alkaline for other perennial legumes, and survives periods of flooding up to 6 weeks. In Canada, it is used in short rotations to improve soil fertility and structure in areas such as the gray wooded soil zones. It grows best where climates are cool and soil moisture is abundant (rain falls over 950 mm or 38 in.). Alsike clover withstands low temperatures better and is less susceptible to winterkill than red clover. Like other clovers, it responds well to phosphates.

Aurora is a cultivar of alsike clover bred at the Agriculture Canada Research Station at Beaverlodge, Alberta. It is superior to other cultivars in forage and seed yield, as well as in winter hardiness. *Tetra* is a tetraploid cultivar bred at the Plant Breeding Institute in Wiebullsholm, Sweden. The increase in ploidy level which the name implies has resulted in *Tetra* giving higher yields than the diploid cultivar from which it was derived. However, it does not outyield *Aurora*.

Red Clover (*Trifolium pratense* L.), also called purple clover, meadow clover, and cowgrass, is a hairy, short-lived perennial (2 to 7 years) found in subarctic, temperate, and subtropical regions. It is second only to alfalfa in the length of time that it has been used as a forage crop. It originated in southeastern Europe and Asia and spread westward in the third and fourth centuries, reaching Spain in 1489. By 1513, Sir Richard Weston was advocating its use, in a mixture with perennial ryegrass, for pastures in Britain. There are reports of its use in Holland, France, and Germany during the sixteenth century. It was one of the first legumes brought by settlers from Europe to North America.

In common with alfalfa, long exposure to agricultural usage in a diversity of climates has led to the emergence of a number of different types of red clover. These are divided into two main groups (single-cut and double-cut red clover) and an intermediate type. The single-cut types have a longer photoperiod requirement for the initiation of floral primordia than do the double-cut types; hence, they are later flowering. Also, they are more winter hardy and larger than the double-cut types and form leafy rosettes with very few flowering heads during the year of establishment. All of the red clovers are cross-pollinated. There are both diploid ($2n = 14$) and tetraploid ($2n = 28$) forms.

When in the rosette stage, red clover is well able to withstand intensive grazing,

FIGURE 4-4
Trifolium hybridum L. Alsike clover.

88 / The Perennial Legumes

FIGURE 4-5
Alsike clover. The leaflets are somewhat rounded or elliptical and have no "watermarks."

since axillary and terminal buds are near ground level. This species may also be used for hay because of its upright habit in the second year of growth. Red clover grows well with a great many different grasses. For example, it is used with smooth bromegrass or timothy, while in the southeastern United States it might be planted as a winter crop in bermudagrass pastures with annual grasses. It is also grown for winter forage production in the southern states with Italian ryegrass, and may be sown into established stands of orchardgrass or Kentucky bluegrass, or in a fescue pasture.

Red clover thrives in humid regions with moderate or cool temperatures. It will not withstand drought, but the single-cut type can tolerate cold winters. It can withstand soil acidity much better than alfalfa, alsike clover, or sweetclover; its yield is reduced only when the soil pH is below 5. In general, however, red clover does not yield as much forage as alfalfa. Over a 3-year period at Beaverlodge, Alberta, it averaged 2.7 T/ha (1.2 tons/acre), while alfalfa averaged 3.3 T/ha (1.5 tons/acre).

Red clover is easily recognized by its hairy leaves, horseshoe-shaped "watermark" (Figure 4-6), and red-veined stipules. In view of the wide variation within the species, the selection of a cultivar well adapted to local conditions is very important. *Kenstar*, a cultivar resistant to southern anthracnose, was developed by the U. S. Department

FIGURE 4-6
Trifolium pratense L. Red clover.

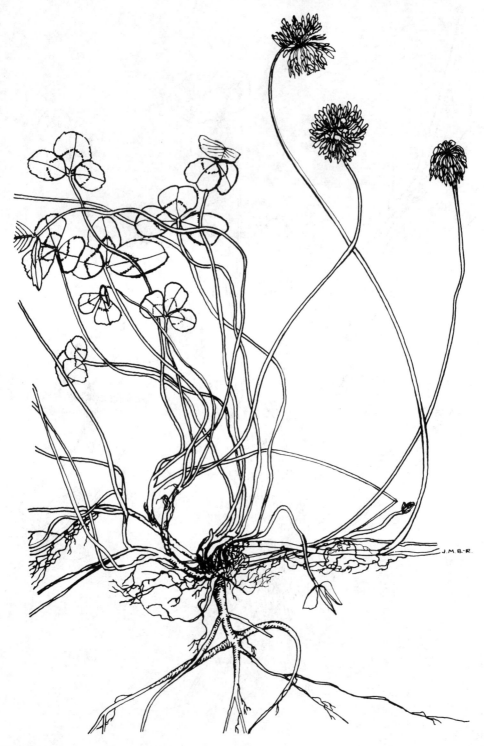

FIGURE 4-7
Trifolium repens L. White clover.

of Agriculture in Kentucky. *Redland* is a cultivar with a wide range of adaptation in the United States.

White clover (*Trifolium repens* L.) is a stoloniferous, white-flowered, long-lived perennial plant (Figure 4-7) which is as widely distributed throughout the world as any other legume. Its early history and spread is not well documented, but its origin is believed to be in the Near East. More recent records indicate it was grown in the Netherlands in 1570 and in Britain in 1694. In the early days of European settlement in the North American continent, its American Indian name was "white man's foot grass."

There are three types of white clover: the small *New York* or *English wild* white clover, intermediate-sized *Louisiana* white clover, and large *ladino* clover. Wild white clover is primarily a pasture plant and is most frequently grazed. It grows well in a mixture with small plants like Kentucky bluegrass. Louisiana and ladino types are grown, for example, with dallisgrass, bermudagrass, and carpetgrass in the southeastern United States. However, careful management is needed for good growth in warm climates, since white clover is more cold than heat resistant and will summer-kill where temperatures are high. It grows best in the humid, cool area of the temperate zone. Where conditions are favorable, white clover will spread under heavy grazing (Figure 4-8). White clover is shallow rooted and produces little growth under dry conditions. It provides palatable, nutritious grazing relished by all classes of livestock.

FIGURE 4-8
White clover.

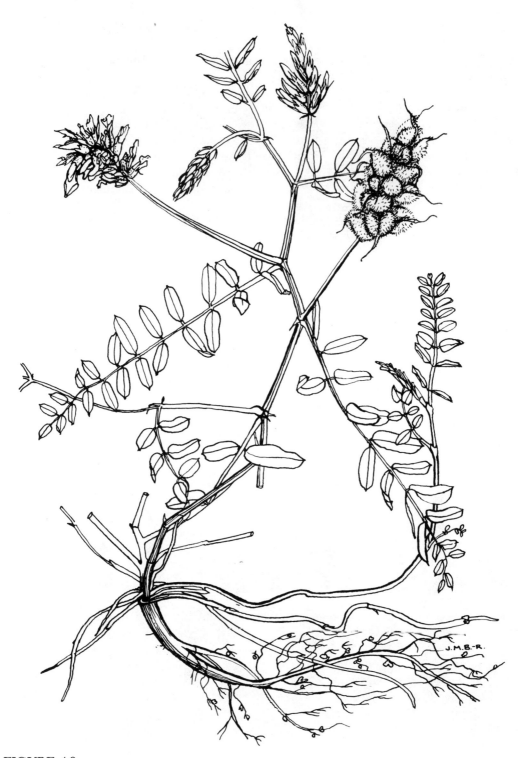

FIGURE 4-9
Astragalus cicer L. Cicer milkvetch.

Species Used in Temperate Areas / 93

FIGURE 4-10
Cicer milkvetch showing leaves, flowers, and the large pod.

Because cultivars of white clover show considerable variation in their photoperiodic responses, care should be taken to obtain locally adapted types. Tests in Alabama have shown *Regal* ladino clover to tolerate wet soil conditions. *Louisiana S-1* and *Nolin* (both of which are intermediate types) have good heat tolerance in the southern United States.

Cicer Milkvetch (*Astragalus cicer* L.)

The genus *Astragalus* contains numerous perennial herbs, few of which are of agricultural value. Cicer milkvetch is an exception. It was used as a pasture legume in Europe before it was introduced to the United States in 1923 and Canada in 1931. It is a rhizomatous, long-lived perennial so vigorous that a single plant may spread to as much as 70 cm (2 ft) in diameter in a single year. The stems are frequently 1.2 m (4 ft) long. The pinnate, compound leaves have from 16 to 22 leaflets (Figure 4-9). The yellow or white flowers are borne in racemes containing 5 to 40 flowers. The plant is cross-pollinated, producing black bladderlike seedpods containing 3 to 12 yellow shiny seeds (Figure 4-10). The seed is larger than those of alfalfa.

Cicer milkvetch is adapted to the Great Plains and western states of the United States and to western Canada. It is usually used for pasture, less frequently for hay, and

FIGURE 4-11
Onobrychis viciaefolia Scop. Sainfoin.

TABLE 4-2
Production and consumption of three legumes,
Lethbridge, Alberta

Legume	Year after establishment:	Yield as a percentage of alfalfa (%)			Consumption as a percentage of production (%)		
		3rd	4th	5th	3rd	4th	5th
Alfalfa		100	100	100	68	65	79
Sainfoin		99	101	95	61	73	88
Cicer milkvetch		70	79	61	59	61	74

(Source: Agriculture Canada Publication 1536)

is also useful in preventing soil erosion. It is extremely winter hardy and survives conditions in the Yukon, where alfalfa, all clovers, and birdsfoot trefoil are killed out after 3 to 5 years. Cicer milkvetch is salt tolerant, drought resistant, able to withstand flooding, and will grow on soils too alkaline or too acid for alfalfa. For farmers growing cicer milkvetch, the supply of seed presents a serious problem. The seed pod becomes very hard when mature, making it difficult to extract the seeds. Not only are seed yields low, but a large number of "hard" seeds, impermeable to water and air, are produced. Even when the seed is scarified, the germination percentage is very low (30% to 36%). The growth rate of the young seedling is slow, and good stands are not always obtained in the first year. Also, when established at Lethbridge, Alberta, neither the forage yield of cicer milkvetch nor its consumption by ewes was as high as that of alfalfa or sainfoin (Table 4-2). The dry matter digestibility (62%) for cicer milkvetch was found to be slightly higher than that for alfalfa (59%). No cases of bloat have been reported for this species. *Oxley,* bred in Alberta, Canada, for high seed yields is used both there and in Montana. In 1970, *Lutana* was released in Montana and Wyoming for use in the United States.

Sainfoin (*Onobrychis* spp.)

While most of the sainfoin grown agriculturally is *Onobrychis viciaefolia* (Figure 4-11), there are also two other similar species in use. The genus originates in central and southern Europe and the temperate parts of Asia, where it has been grown as a forage legume for the last 500 years. It was introduced into North America about 120 years ago, but it was not until the cultivar *Eski* was developed from Turkish material at the Montana Experiment Station in 1960 that it was used agriculturally.

Sainfoin is deep rooted and very drought resistant, provided the annual rainfall is 300 mm (12 in.) or over. It yields best on deep, well-drained soils and will not withstand

FIGURE 4-12
Melilotus officinalis L. Yellow sweetclover.

wet soils or high water tables. Tests show that it will not tolerate saline soils and that it is not as winter hardy as the locally recommended cultivars of alfalfa.

Sainfoin grows taller than alfalfa, its stem is hollow, there are many leaflets (like a vetch), and its flowers, which are pink, are borne on a raceme (Figure 4-11). The "seed" used to establish this crop is, in fact, a pod which contains a single seed. Even without the pod, the true seed is large (for a legume); there are only 62,000/kg (28,000/lb). Seeding rates are high, about 20 kg/ha (18 lb/acre). Compared with alfalfa, forage dry-matter yields of sainfoin are about 20% lower under dry land conditions and may be 30% or more lower in irrigated areas.

The primary reason for using sainfoin is that, throughout their long history, *Onobrychis* species have never been known to cause bloat nor is it attacked by alfalfa weevil. It is highly palatable to both sheep and cattle (see Table 4-2), being preferred over alfalfa. It may be grazed or used for hay, either alone or in mixtures with grasses. It grows well with Russian wildrye and crested wheatgrass. Under irrigation, sainfoin is shorter-lived than alfalfa, but rotational grazing has been shown to prolong its life.

Melrose was released in 1969 for use in Canada and, together with *Eski*, is now used throughout the western parts of North America.

Sweetclover (*Melilotus* spp.)

There are two species in this genus which are widely grown in North America. They are *M. alba* Deor., white-flowered sweetclover, and *M. officinalis* L., yellow-flowered sweetclover (Figure 4-12). The genus is native to Asia Minor and was introduced to North America in about 1700. There are reports of its agricultural use in Virginia in 1739, New England in 1785, and Alabama in 1856. Prior to the nineteenth century, it was considered to be a weed rather than a forage crop. The plant produces a large amount of seed and frequently establishes itself on roadsides.

Sweetclover is an excellent legume for soil improvement because of its high nitrogen fixation rate. It has branched, deep taproots which penetrate into the subsoil. If it is plowed into the soil when the plants are 15 cm (5 in.) high, the roots and the top growth will decompose to release nutrients and provide organic matter to improve soil structure. In the Corn Belt of the United States, it is used in short rotations to improve soil conditions. It will grow in many soil types, provided the soil is well drained and not acid (above pH 6).

Sweetclover is adapted to most of the environmental conditions found in North America. It is grown in the south, yielding well on the black soils of Texas and the limestone soils of Alabama and Mississippi. It is extremely drought resistant and grows well on those parts of prairies and the Great Plains where the annual rainfall is about 400 mm (16 in.). It has shown itself to be very winter hardy.

The leaves and their leaflets are rather like those of alfalfa, but, in contrast with alfalfa, the leaflets are toothed around the whole of their margins (Figures 4-1 and 4-12). The sweetclover plant is many branched and becomes woody as it matures. It may be grown as an annual, but is usually a biennial. The percentage of crude protein declines from 21% in young plants to 13% at the hay stages; at the same two stages, the percentage crude fiber is 19% and 36%, respectively. At the hay stage, sweetclover

FIGURE 4-13
Vicia sativa L. Common vetch.

is sweet scented because it contains coumarin. Coumarin may be reduced to dicoumarol, a substance that will prevent blood from clotting and so cause "bleeding" disease in livestock. The sweetclover antiquality component is discussed in Chapter 10.

Sweetclover may be used in pasture and, less often, for hay and silage; it is also an excellent source of nectar, providing good yields of high-quality honey. For sheep and cattle, the danger of bloat is less with sweetclover than it is with the true clovers or alfalfa. Sweetclover is not a good crop for hay. The stems are thick and slow to dry; the leaves are fine, dry quickly, and are likely to become brittle. The use of a conditioner attachment when hay making may solve this problem and reduce the danger of bleeding disease (see Chapter 10). In general, sweetclover is neither as high yielding nor as palatable as alfalfa. For example, in South Dakota, the weight gains of lambs when fed prairie hay (native species), sweetclover hay, and alfalfa hay averaged 0.16, 0.20, and 0.25 kg (0.35, 0.44, and 0.55 lb) per day, respectively.

Hubam is an annual white sweetclover widely used in the lower south and southwestern United States. *Madrid*, also recommended for southern areas, is one of the better yielding yellow-colored cultivars. In Canada, *Polara* is grown. It is a white-flowered biennial variety which was licensed in 1970. Because it has a low coumarin content, it will not cause bleeding disease. *Yukon*, developed in Saskatoon, Saskatchewan, is a yellow sweetclover cultivar which is suitable for the north, being a winter-hardy strain of *Madrid*.

Vetches (*Vicia* spp.)

There are about 150 members of this genus, of which 11 are important in North American agriculture. The genus is native to Europe, Asia, and parts of Africa. Vetches are weak-stemmed climbing plants with pinnate leaves and terminal tendrils (Figure 4-13). The inflorescences are racemes. In general, they are not very winter hardy and are grown as summer annuals in areas with a cool climate, and as winter annuals in warmer areas. They are used for grazing when grown alone or with a small-grain crop. When grown in mixtures, they may be used for hay or, more often, for silage.

Hairy vetch, *V. villosa*, is an annual which sometimes grows as a biennial. It is more winter hardy than other vetch species and is adapted to somewhat acid sandy soils. Common vetch or tare, *V. sativa* (Figure 4-13), is adapted to free-draining and fertile soils. Both species are commonly used as winter annuals in southern United States.

WARM-CLIMATE LEGUMES

Some of the legumes discussed previously (alfalfa, birdsfoot trefoil, sweetclover, and white clover) may be used under both temperate and subtropic conditions. Here we will consider legumes that grow only in regions with warm climates. The warm-climate grasses are frequently of poor quality; they will support only dry cows or stock on maintenance rations unless they are heavily fertilized and closely grazed. When beef cows and calves, dairy cattle, or stocker cattle are being grazed on summer pastures, the presence of a legume will provide an important addition to the nutrient value of the herbage. The legume is important because both temperate and tropical grasses show a high negative correlation ($r = -0.79$) between dry matter digestibility and temperatures at the time of maximum vegetative growth. This relationship is not found in legumes. Neither white clover nor siratro (a tropical species) has different dry-matter digestibilities between summer and winter. Also, as grasses age, they become more fibrous and their digestibility decreases. Most of the temperate legumes follow the same pattern, but there are some warm-season legumes which maintain a high digestibility and voluntary intake. Cowpea and Rongai lablab are examples of species that behave in this way. Thus, warm-season legumes play an important role in increasing and stabilizing forage quality.

Legumes also make a valuable contribution to both quality and production in warm climates whose growing season terminates in a drought or which are subject to periods of drought during the season. At such times, many legumes remain green and continue to produce. This is because most legumes have a taproot which will go deeper into the soil than the shallow fibrous grass roots. Also, the fleshy root crown found in legumes provides for the storage of nutrients and water.

For the humid southeastern United States, *Sericea*, *Lespedeza*, and *Kudzu* are species which make an important contribution to forage production between late April and the beginning of September. For example, a mixture of coastal bermudagrass and *Sericea* or *Lespedeza* is well adapted to the sandhill areas of southeastern North Carolina. In the lower south of the United States, annual warm-season legumes are frequently used to overseed warm-season grasses. Under these circumstances, ryegrass, with crimson or arrowleaf clover, is used for forage production in winter months (November to June), giving peak production in April and May.

The value of legumes in conjunction with high-yielding grasses in the tropics is well illustrated by the yields of mixtures of ladino clover and orchardgrass or alfalfa and grass. In the humid southern states of America, such species can yield between 6,000 and 9,000 kg/ha (5,340 and 8,010 lb/acre) of dry matter. To obtain this production level from grass alone, the plants would require a fertilizer dressing of 100 to 200 kg/ha of nitrogen (89 to 178 lb/acre). The presence of a legume in the pasture also increases the digestibility of the herbage and improves the balance of its mineral content. A nitrogenous fertilizer would not provide these additional advantages.

The importance of legumes in summer pastures in the tropics and semitropical areas of the world cannot be overstressed. Given good management practices, the species discussed in the following pages can greatly improve livestock production, even if they do not provide a fully satisfactory solution to the problem.

Arrowleaf Clover (*Trifolium vesiculosum*)

This true clover is a self-reseeding winter annual which gives a high yield of good-quality forage in the late spring. It matures in May or June, about 2 months after crimson clover, and may be used for grazing, hay, or silage. It is less acid tolerant than crimson clover, growing best in soils with a pH between 6 and 7. This species has good drought tolerance and starts growing rapidly after rain.

Arrowleaf clover is usually planted with a small grain (rye or wheat), ryegrass, bermudagrass, or bahiagrass. It is also used to overseed existing bermudagrass and fescue pastures. This is accomplished by discing in 3 to 4 kg (6 to 9 lb) of seed per hectare in late August or September. Growth during the fall and winter months is limited, but is rapid in late spring. Its growth habit makes arrowleaf clover suitable for hay, silage, and cool-season grazing.

The seed of this plant contains up to 80% "hard" seed, which may take up to 4 years after planting to germinate. For a new seeding, 2-year-old seed, which has been scarified, should be used. In established stands, arrowleaf clover will produce enough hard seed to maintain or reestablish itself. Heavy grass growth from the summer should be removed, as it retards germination. A second year's volunteer crop is usually produced early in the fall and is higher yielding than a new seeding.

The forage quality of arrowleaf clover is high. The digestible dry-matter content is 80% at the peak of vegetative production and 70% at the early bloom stage. Animal gains are good. In a trial conducted in Alabama, feeder steers gained 1 kg (2.2 lb) per day when grazing a pasture containing ryegrass and arrowleaf clover.

Crimson Clover *(Trifolium nicaenatum)*

This species, growing 30 to 90 cm (1 to 3 ft) tall, is a hairy, upright winter annual with a small taproot. Its stem and leaf are rather like those of red clover, but the flower color is different, being a brilliant crimson. The flower opens successively from the bottom. Crimson clover originates in Europe and is still grown in France, Hungary, Italy, and southern England. It was introduced into the United States in 1819 and is now an important legume in the central and eastern states, as well as in Kentucky, southern Ohio, Missouri, the coastal areas of California, and states along the Gulf of Mexico. The plant is moderately winter hardy, but not very drought resistant. It can withstand a high lime content and poorly drained soils. It grows best on well-drained, fertile soils in cool, humid climates with an annual precipitation of about 875 mm (35 in.). It will tolerate acid soil conditions better than white or red clover and alfalfa or sweet clover.

Crimson clover is used for pasture, hay, and soil improvement. It is eaten with relish by all classes of stock and is commonly seeded with a small-grain crop or with ryegrass. The yield of crimson clover will be less than that of white or red clover when grown in a mixture with coastal bermudagrass. Crimson clover is normally seeded between August and October in the upper south of the United States and about 1 month later in the lower south.

There are a number of cultivars of crimson clover *(Dixie, Talladega,* and *Kentucky)* which reseed themselves. These cultivars contain 30% to 75% hard seed. The ability to reseed is not fully expressed in common crimson clover.

Kudzu (*Pueraria lobata* **Willd.**)

This species is a long-lived, coarse, vigorously growing perennial which is indigenous to Japan and China. It has now been introduced to most tropical countries in the world, reaching the southern United States in 1876. It has very heavy stoloniferous stems, petioles, and pods. The vines are frequently 20 m (66 ft) in length. There is a hairless mutant which is reported to be more palatable than the normal type. The plant has many stolons which intertwine and cling tenaciously to the soil by rooting at both nodes and internodes (Figure 4-14).

Kudzu is adapted to well-drained heavy and sandy soils. It exhibits no winter hardiness and grows only where the vines are not killed back to the crown in winter. It has good drought resistance, remains green for 2 months following rain, and is most frequently used on sandy soils. The seeds ripen very unevenly, and seed set is poor, making harvest difficult and expensive. When available, the seed is "hard" and has to be scarified with either hot water or sulfuric acid. For these reasons, this species is frequently established from crowns or naturally established seedlings. Under either circumstance, it may take 2 or 3 years to establish a full ground cover, since the plants grow slowly for the first 3 to 4 months. For seed production, kudzu has to be grown on trellises, as the plant produces few flowers when allowed to grow on the ground.

Kudzu may be used for green manure, as a cover crop in rubber or oil palm plantations, and for grazing, hay, or silage. It will not take heavy or continuous grazing, but is highly palatable and resistant to trampling when rotationally grazed. The hay equals alfalfa in quality and yield, having a higher leaf to stem ratio (about 1:1) than alfalfa. It is regarded as a weed in southeast United States.

Annual and Perennial *Lespedeza* Species

Some members of this genus are indigenous to Central America, while others were first introduced in 1840 into the United States from Asia and Australia. There are two annual species in agricultural use in the United States: Kobe lespedeza *(L. striata)* and Korean lespedeza *(L. stipulacea)*. These two species are very drought resistant and do well on poor, sandy soils. They will thrive where soil conditions are acid (pH 5.5 to 6.5). They can be used to produce a good-quality hay, provided they are raked and baled soon after cutting to avoid leaf shed. Improved perennial cultivars and better management practices have led to annual lespedeza being replaced by species such as arrowleaf clover.

Perennial lespedeza, however, is still common in the agriculture of the southern United States. Sericea lespedeza *(L. cuneata)* is a long-lived perennial with a deep taproot. It was introduced into the United States from the Far East in 1900. It is a woody, branched plant rather like alfalfa (Figure 4-15) and is grown in the Piedmont section of the upper south.

Sericea lespedeza may be used for soil protection, hay, and summer pasture. It is an excellent soil builder, yielding well on acid soils of low fertility and is used in this way in Texas, Oklahoma, and Kansas. It responds well to lime and to fertilizers. Also, lespedeza species have a high tannin content, which is associated with low palatability and low digestibility if not harvested or grazed early.

Subterranean Clover *(Trifolium subterraneum)*

This species, which is frequently called "subclover," is an annual, self-seeding true clover, widely grown in Australia and New Zealand for sheep and cattle. It is a native

FIGURE 4-14
Pueraria lobata Willd. Kudzu.

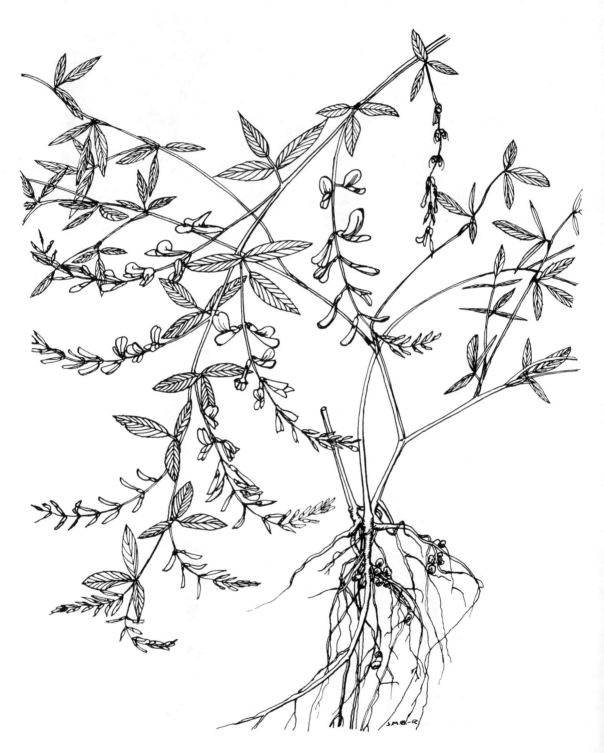

FIGURE 4-15
Lespedeza cuneata L. Sericea lespedeza.

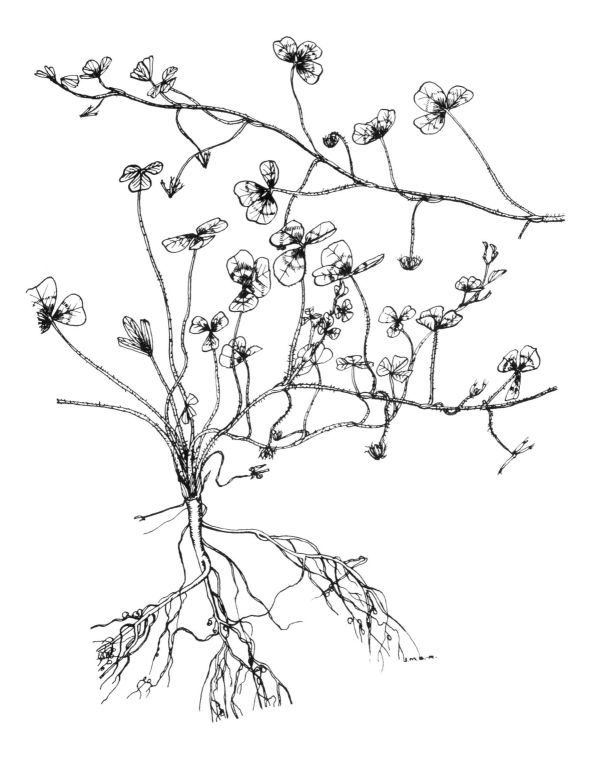

FIGURE 4-16
Trifolium subterraneum. Subterranean clover.

of the European Mediterranean area, has some winter hardiness, and has foliage which can withstand temperatures as low as −7°C (19.4°F). It is also somewhat acid tolerant, but will not survive on poorly drained soils. The plant is day-length neutral (Figure 4-16).

In the flowering stage, the plant is unpalatable to all classes of stock and, as a result, self-seeding occurs frequently. *Mt. Barker* is the cultivar that is best adapted to the southeast United States. It has also given satisfactory cover when used for the revegetation of disturbed mountain areas in Washington. Fescue, bermudagrass, and ryegrass are suitable species to grow in a mixture with subterranean clover. The forage yield from subterranean clover is then about the same as that from crimson or arrowleaf cover. Subterranean clover is also used with small-grain crops or ryegrass to provide fall grazing. However, it is usually most productive in the spring. Seed harvest is complicated because the burs which contain the seed are below the soil surface. The plant is pulled up, dried, and the burs are then removed by a pickup combine.

Reports from Australia indicate that reproductive disorders are common in ewes grazing subterranean clover. This was believed to be the outcome of the overstimulation of the animal's reproductive system by estrogens produced by subterranean clover and some other clovers. Subterranean clover contains estrogens during the entire growth period, but the content is low when the plant is dry or mature. The causal compound is genistein, an isoflavon. Similar substances, formononetin and biochanin A, are present in other legumes and some grasses. Isoflavons are not as active as the coumestan compounds discussed in the section on flavonoids in Chapter 10. The estrogenic substance in subterranean clover appears to be associated with a failure of sperm movement in the ewe's genital tract, which results in infertility. The severity of this condition increases when potash, nitrogen, and sulfur are deficient in the soil or when levels of copper are low.

FURTHER READING

Elliot, C. R., and J. A. Bolton,   1970, Licenced Varieties of Cultivated Legumes, Agriculture Canada Publication 1405, Ottawa.

Elliot, C. R., and P. Pankew,   1972, Alsike Clover, Agriculture Canada Publication 1264, Ottawa.

Fernald, M. L.,   1950, Gray's Manual of Botany, American Book Co., New York.

Goplen, B. P., and A. T. A. Gross,   1977, Sweetclover Production in Western Canada, Agriculture Canada Publication 1613, Ottawa.

Gunn, C. R.,   1971, Seeds of native and naturalized vetches of North America, USDA Handbook 392.

Hafenrichter, A. L., J. G. Schwendimen, H. L. Harris, R. S. MacLaughlin, and H. W. Miller,   1968, Grasses and legumes for soil conservation in the Pacific Northwest and Great Basin states, USDA Handbook 339.

Hanna, M. R., D. A. Cooke, S. Smoliak, B. P. Goplen, and D. B. Wilson, 1977, Sainfoin for Western Canada, Agriculture Canada Publication 1470, Ottawa.

Hanson, C. H., ed., 1972, Alfalfa Science and Technology, American Society of Agronomy, Madison, Wisconsin.

Heinricks, D. H., 1969, Alfalfa in Canada, Agriculture Canada Publication 1377, Ottawa.

Henson, P. R., 1957, The lespedezas I, Advan. Agron. 9:113-22.

Isley, D., 1951, The legumes of the north central United States. Iowa State Coll. J. Sci. 25:439-82.

Johnston, A., S. Smoliak, M. R. Hanna, and R. Hanna, 1975, Cicer Milkvetch for Western Canada, Agriculture Canada Publication 1536, Ottawa.

Pankew, P., C. R. Elliot, L. P. Folkins, and H. Baenziger, 1977, Red Clover Production, Agriculture Canada Publication 1614, Ottawa.

Romm, H. J., 1953, The development and structure of the vegetative and reproductive organs of kudzu, Iowa State Coll. J. Sci. 27:407–19.

Smith, W. K., and H. J. Gorz, 1965, Sweet clover improvement. Adv. in Agronomy 7:163–231.

U S Department of Agriculture, 1971, Growing crimson clover, Leaflet 482, pp. 1–10.

chapter 5

Annual Forages

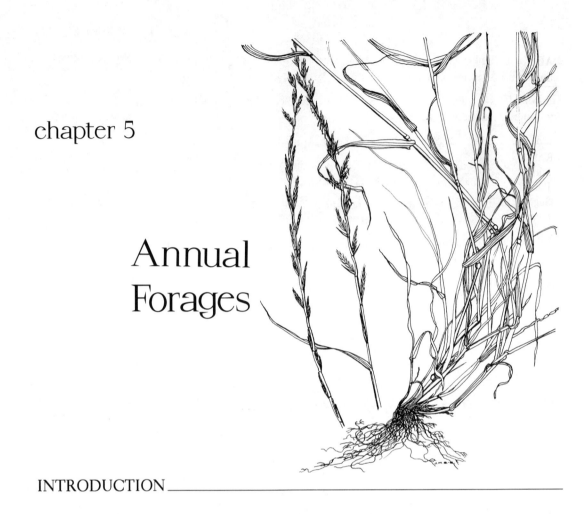

INTRODUCTION

There are a large number of annual plant species whose vegetative parts, harvested before the seed or grain is fully formed, make excellent forage. Annual plants, unlike perennials, store almost all their carbohydrate reserve material in the seed. To obtain bulky, fibrous material which is also digestible and nutritious, annual plants should be harvested before excessive lignification takes place and before carbohydrate reserve material is moved from the leaves and stems into the seed. For the small-grain crops, this is when the grain is in the milk or early dough stage.

Annual forage may be used agriculturally under all climatic conditions. For summer use in the tropics, there are grasses (e.g., sudangrass), grain crops (e.g., corn, sorghum, and cattail or pearl millet), legumes (e.g., lespedeza), and grain legumes (e.g., soybeans and peanuts). In the humid south of the United States, it is possible to maintain stock throughout the year by using annual forages. Pearl millet or sudangrass provides summer grazing, soybeans give good-quality forage in large amounts in late summer, velvet beans are productive in the fall, and winter grazing frequently comes from oats, wheat, barley, rye or ryegrass, and arrowleaf clover, vetch, or crimson clover.

Winter annuals in subtropical to warm areas are used in two ways. First, where winter temperatures are satisfactory for vegetative growth, they are seeded into dor-

TABLE 5-1
Annual forages and their use

	Warm areas		Temperate areas	
			Fall sown	
	Summer use	Winter use	for spring use	Spring sown
GRASSES				
Sudangrass	x			x
Italian ryegrass		x		x
Sugarcane	x			
Sorghum–sudangrass hybrid	x			x
LARGE-GRAIN CROPS				
Corn	x			x
Sorghum	x			x
Cattail or pearl millet	x			x
SMALL-GRAIN CROPS				
Barley				x
Rye		x	x	x
Oats		x		x
Wheat		x	x	x
Triticale		x		x
TROPICAL LEGUMES				
Arrowleaf clover		x		
Crimson clover		x		
Cowpeas	x			
Lespedeza	x	x		x
Soybean	x	x		
Sweetclover		x		x
Velvetbeans	x			
Vetches	x	x		
TEMPERATE LEGUMES				
Beans				x
Lupines				x
Peas				x
Subterranean clover		x		
OTHER CROP SPECIES				
Kale				x
Rape				x
Sugar beet				x
Sunflower				x

mant permanent pastures composed of warm-season species. For example, ryegrass or crimson clover, or both, are frequently sown into established bermudagrass stands. Second, where there are long, cold periods in the winter, with temperatures at $-5°C$ (23°F) or somewhat below, winter cereal crops such as rye or wheat are sown in the fall for early grazing the following spring. Depending on the summer temperature, many of the species that are grown as winter annuals in warm areas are also used as summer annuals in more temperate climates. Table 5-1 lists the main annual forage crops in agricultural use and indicates their areas of climatic adaptation.

WHY ANNUALS?

There are five reasons why annual forages are popular. First, temperate climate annuals may be used in the cool winters when warm-season perennials are dormant. For example, bermudagrass may be oversown with oats when it becomes unproductive in the fall. Second, they introduce flexibility into the farming system. If some of the land is not productive because it is the establishment year for a new pasture, or if a farmer has decided to increase the number of animals on the farm, then annual forages may be used to provide extra feed quickly. These crops are frequently able to provide forage material in less than 4 weeks. In regions with cold winters, annual forage crops may be grazed well into the fall without regard for carbohydrate reserve levels (see Chapters 7 and 8) or concern for winterkill. Third, where a well-adapted annual forage is used, it can give a very high yield indeed. Enhanced productivity will offset the extra expense of annual establishment. Increases in grain prices in recent years have led many farmers to use annual forages to make silage for overwintering their livestock. Fourth, compared with perennials, many of the species listed in Table 5-1 are easily established. The small-grain crops, for example, have a much larger seed than the grasses and do not require the same care in seedbed preparation or in planting depth. Finally, most annual forages are palatable, well liked by stock, and give good animal-weight gains.

As in many other agricultural situations, the choice between an annual and a perennial forage is determined by economic considerations. What is economically most expedient will vary with market prices and farming circumstances; a wise choice is not easily made. At the University of Alberta farm, a smooth bromegrass variety trial established in 1976 gave dry-matter yields of 12,499, 5,734, and 4,371 kg/ha (11,124, 5,103, and 3,890 lb/acre) for the subsequent 3 years and an average annual yield over the 3-year period of 7,534 kg/ha (6,705 lb/acre). These figures may be compared with the best annual forage yields, which vary between 7,500 and 8,000 kg/ha (6,675 and 7,120 lb/acre), for this same area. Here, it is evident that the additional cost of establishing the annual offers little, if any, economic advantage over a perennial crop. Consideration of this type would not apply to the southern United States, where the addition of a winter annual would give year-round forage production and always result in a financial gain.

High yields and economic advantage are usually the outcome of close adaption of the plant to local soil and climatic conditions. Yields from trials conducted in Alberta are presented in the form of regression lines in Figure 5-1. The "trial mean" may be regarded as an index of productivity so that an identical yield level is to be expected where the lines cross. In Figure 5-1, the regression lines cross at about 5,300 kg/ha (4,717 lb/acre). At higher productivity levels, oats give the more favorable yield, and at lower productivity levels, barley is more productive. In these trials, the higher-

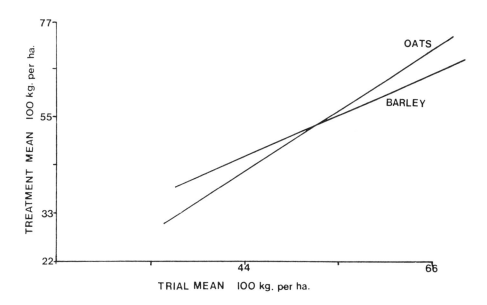

FIGURE 5-1
Relative yields of oats and barley, grown for forage, at different fertility levels (trial means): kg/ha × 0.89 = lb/acre. (Source: P. D. Walton, 1975, Can. J. Plant Sci. 55)

yielding areas had a longer growing season, a higher rainfall (450 mm or 18 in.), and a more fertile soil. It is evident that crop selections for a particular location should be made with great care.

USES

Annual forage plants may be used for hay or silage, chopped and fed green, or grazed. In the southern United States, small-grain crops may be grazed early in their growth and then allowed to mature to produce a grain crop. Where a small-grain crop is used for hay, seeding rates are normally high (60 to 90 kg/ha or 53 to 80 lb/acre) and harvesting takes place in the late milk or early dough stages. If the hay is intended for horses, the crop may be harvested somewhat later so that the grain has longer to develop. They should be finely chopped and well packed in the silo.

Work at Melfort, Saskatchewan, has shown that small-grain crops, when grazed, give good weight gains with both cattle and sheep. The seeding rates used to establish the pastures are high. The pasture height when the animals start grazing is critical; it should be about 25 cm (9 in.). Lower heights result in younger herbage being grazed. Young, green oat herbage acts as a laxative, causing digestive upsets and scouring; taller, older, vegetation, however, leads to high trampling losses. Rotational grazing helps to reduce trampling losses, which, even under the most favorable circumstances, are high (about 20%). When annual crop pastures are seeded for rotational grazing, crop establishment is on sequential dates so that as each field reaches a height of 25 cm (9 in.) the animals are placed in that pasture.

The long growing season in the southeastern United States permits a wide range of forage-production systems which use both warm- and cool-season annuals. These systems provide either year-round grazing, or call for only a little supplementary conserved fodder or grain in December and January, for high animal production. The most frequently used grass is tall fescue, which has a long winter growing season. This species is usually sown with a legume or another grass to improve the quality of the forage. The most commonly used legumes are arrowleaf clover, subterranean clover, and red or crimson clover. These annual clovers are usually seeded early in the fall while the tall fescue plants are still summer dormant.

High-quality winter annual pastures are usually established by seeding into land which has been cultivated to produce a well-prepared seedbed. The species used in such pastures are ryegrass or small-grain crops such as wheat, rye, or oats. In the humid south of the United States, rye is the most common choice of small grains for sandy soils. Wheat and oats are used on heavier soils. In general, ryegrass is regarded as the best annual pasture grass for the humid south. Small-grain crops are also used to overseed dormant bermudagrass pastures.

Winter pastures involve some cost in establishment and, for the best economic returns, should be of high quality, and should be used primarily for dairy cattle and sheep or animals that will be sold directly off the pasture (see Table 5-2). Summer annual pastures should also be of high quality and be utilized by the most productive classes of stock.

Of the two types of annual pasture (summer and winter), the summer pastures are the most difficult to manage well. To achieve the appropriate high quality, high dressings of nitrogen fertilizer are used. For summer pastures, sudangrass or sorghum–sudangrass hybrids are most commonly grown. Pearl millet is well adapted to dry

TABLE 5–2
Animal weight gains from overseeded bermudagrass pastures, Auburn

| | Animal weight gain | | | |
| | Cows | | Calves | |
Winter annual species	kg	lb	kg	lb
Ryegrass, arrowleaf, and crimson clover	84	185	236	519
Arrowleaf and crimson clover	110	242	184	404
Ryegrass	93	204	188	413
None	62	136	118	259

(Source: Hoveland et al., 1978, Agron. J. 70. By permission of The American Society of Agronomy, Inc.)

TABLE 5-3
Effect of overseeding a sorghum–sudangrass hybrid in a tall fescue pasture, in southeastern United States, mean of 3 years

	Summer yield	Winter yield	Total yield
METRIC	Dry Matter (kg/ha)		
Tall fescue only	5,073	2,332	7,405
Tall fescue and hybrid	4,876	2,418	7,294
50% Tall fescue killed and hybrid	7,207	2,082	9,291
ENGLISH	Dry matter (lb/acre)		
Tall fescue only	4,515	2,076	6,590
Tall fescue and hybrid	4,340	2,152	6,491
50% Tall fescue killed and hybrid	6,414	1,852	8,269

(Source: Belesky et al., 1981, Agron. J. 73. By permission of The American Society of Agronomy, Inc.)

conditions on sandy or medium-textured soils. Foxtail millet is also a fast-growing, popular summer annual.

Annual summer pasture plants may be sown in a dormant tall fescue pasture to increase the summer production. Unless a high-yielding species (like a sorghum–sudangrass hybrid) is used, the total annual production may not increase, since the yield of tall fescue will be reduced (see Table 5-3). For this reason, the annual summer forages are usually grown alone. An exception would be to plant annual lespedeza or another annual legume crop to improve quality in a bermudagrass pasture.

To take advantage of areas like the southeast of the United States, where year-round grazing with little or no supplementary feed is possible, the farmer needs a sound grazing system (or plan) and a number of the annual forages discussed previously. Such a plan should be simple for ease of implementation and should provide a rest period for the different types of pasture. This rest period should be long enough and at an appropriate time to benefit the various plant species. The time of utilization should be such that it will be most beneficial to the animals. For example, bermudagrass can be best used by a cow with a calf in early spring. From June onward, dry cows may be maintained satisfactorily on such a pasture, but calf weight gains would decrease. One large herd calls for less labor to move than several small ones. Livestock movement should be timed to coincide with the time of spraying, drenching, testing, or weaning, when the animals would be moved anyway. Distances should be as short as possible. Above all, the grazing system should be flexible. The "unexpected" *will* happen. It always does!

TABLE 5-4
Mean yields of dry matter and crude protein for annual forage crops grown at the University of Alberta farm, 1975 to 1978

	Yield of dry matter		Crude protein	Yield of crude protein	
	kg/ha	lb/acre	%	kg/ha	lb/acre
Oats	5,062	4,505	9.0	450	401
Oats with 25% peas	4,651	4,139	11.9	547	487
Oat–peas in equal parts	4,273	3,803	12.8	536	477

(Source: P. D. Walton, 1975, Can. J. Plant Sci. 55)

MIXTURES

As with perennial forages, annual forage crops are frequently grown in mixtures (Chapter 6). On the University of Alberta farm, a mix of oats or barley with peas gives good yields. However, the addition of peas to the small-grain crop reduces the total yield of dry matter and increases the yield of crude protein (Table 5-4). The choice between oats alone and an oats–pea mixture, in which 25% of the plant population consists of peas, would depend on the purpose for which the forage material was to be used (e.g., dry-cow maintenance or milk production). This choice is essentially between quality and quantity and is one which is frequently encountered in farming practice.

Production figures from the University of Alberta farm show that mixing two species at the same population within a row gives higher yields than alternate row seeding (Table 5-5). This is in contrast with results obtained from perennial forage species (Chapter 6).

CROP RESIDUES

The straw or chaff of row crops grown for grain (e.g., corn, sorghum, rice, wheat, barley, or oats) can be used for grazing after the heads or cobs have been removed. Sugarcane bagasse or sugar-beet tops are also available where these crops are grown. Row-crop residues are low in protein and contain much cellulose, usually highly lignified. Because straw and chaff are frequently exposed to sun and rain, soluble, and hence readily digestible, substances are leached out. The vitamin content and the presence of substances such as carotene are negligible. The mineral content is also frequently low. Such material is unsuitable for high-producing animals and needs to be supplemented. Where straw is to be grazed in the field, it may be supplemented by undersowing the grain crop with a legume. At the best, however, crop residues provide only maintenance rations for overwintering mature cattle.

TABLE 5-5
Yield of dry matter from annual forage crop mixtures grown in the same and alternate rows

	Yield of dry matter			
	Same row		Alternate rows	
	kg/ha	lb/acre	kg/ha	lb/acre
Oats and peas	6,152	5,475	5,948	5,294
Oats and rapeseed	4,427	4,940	4,012	5,370
Barley and peas	6,651	5,919	6,038	5,374
Barley and rapeseed	6,797	6,049	6,329	5,633
Barley and soybeans	6,052	5,386	4,928	4,386
Wheat and soybeans	4,790	4,263	4,057	3,611
Mean	5,811	5,171	5,218	4,644

Least significant differences (5% level) = 412 kg/ha (367 lb/acre)

(Source: P. D. Walton, 1975, Can. J. Plant Sci. 55)

Crop residues do, however, provide feeds that are worth more than their net energy values indicate. This is because heat produced during their metabolism can be used to maintain body temperature. These substances are also of value as dilutants for high-concentrate rations, since some bulky feed is needed for all classes of stock.

As explained in Chapter 9, processing low-quality material will improve both intake and digestibility. This may be done in two ways, either by chopping or grinding the crop residue or by treating it with chemicals. The chemical treatment involves soaking the crop residue in an alkali solution for 1 to 3 days. Sodium hydroxide is the most commonly applied substance, but acids and peroxides are also used. Work in Oregon has shown that grinding, spraying with sodium hydroxide, and then spraying with propionic acid will double the intake of barley straw and more than double the digestible organic matter. Other studies, also in Oregon, show that spraying with sodium hydroxide increases the daily gain for steers from 1.32 to 1.43 kg (2.91 to 3.15 lb).

The alkali treatment of the plant material causes the cell walls to swell, with some degeneration of the chemical bonds between the lignin and the cellulose. Where protein is present, the digestibility is depressed. In most crop residues, protein levels are so low that this is of little concern.

CROP SPECIES USED AS ANNUAL FORAGES

The crop species set out in Table 5-1 are normally classified into six groups: large-grained cereal crops, small-grained cereal crops, annual grasses, tropical legumes, temperate legumes, and a diverse, miscellaneous group of succulent fodders.

Large-Grain Crops

The most widely used and most productive of the large-grain species is corn *(Zea mays)*. Over 4.5 million ha (11.25 million acres) are grown annually in North America. It is used as a summer forage in Canada and the northern United States, as well as in Mexico, South America, and the southern United States. In all areas, it is most frequently used for silage, but may also be grazed. When made into silage, corn does not require additives, since the herbage is high in sugars and other carbohydrates. In southern Alberta, under irrigation, yields of up to 8 tonnes of dry matter per hectare (3.6 tons/acre) have been reported, while in regions with tropical or semitropical climates, yields of up to 100 to 200 tonnes of fresh fodder per hectare (45 to 90 tons/acre) have been reported.

Corn is a cross-pollinating diploid ($2n = 20$) in which the male spikelets (the tassel) form a separate inflorescence from those of the female. The cob, which is a spike, is formed in the leaf axis. Hybrid seed, obtained from two or more selfed lines, is usually grown. When it is intended for fodder, corn is frequently sown with cowpeas or soybeans. Fertilizer applications are essential to obtain high yields from the hybrid seed. The crop is harvested when dry matter has accumulated in the cob and growth is well advanced. The grain is then at the "milk" or "dough" stage, and digestibility values are high (see Table 5-6). Where crude protein content is important, the plant is cut when the tassel is newly formed. If planted at high seed rates, corn can be grazed or fed green. It is, however, one of the best crops for making silage and is frequently used for that purpose. The plants are chopped into pieces 6 to 20 cm (2 to 7 in.) long and compressed to prevent air penetrating the material. Good silage, with a low pH, is usually produced under these circumstances.

Grain sorghums *(Sorghum vulgare)* also make good silage and are used as temporary pastures in the southern United States. They produce good yields, but do not outyield corn. The nutrient value of sorghum silage is increased by cutting in the late dough stage. The whole plant is sometimes cut and shocked, or bedded, for winter feed. It is used in the natural state, but improved animal-weight gains may be obtained by

TABLE 5-6
Yield and digestibility values from corn trials grown in Georgia

Plant part	Dry matter %	In vitro dry matter digestibility %
Stalk	64.2	27.3
Sheath	65.9	6.9
Leaf	64.8	15.7
Ear	79.6	40.1
Husk	63.6	10.0

(Source: Cavines and Thomas, 1980, Agron. J. 72. By permission of The American Society of Agronomy, Inc.)

grinding (see Chapter 9). Grinding is especially beneficial for sorghum, because, if the seed is mature, it passes through the animal undigested.

When grazing sorghum, it is desirable to fence and graze small sections one at a time. Otherwise, cattle will select seedheads and leave much of the remainder of the plant. Under these circumstances, the grazing period for the whole field is shorter, the regrowth is of poorer quality, and the yield is lower. "Sorgo" is a *Sorghum* which gives a low seed yield, but which provides very nutritious forage since it retains a large amount of "sugar" in its stem. Since grain sorghum is subject to weevil damage in the field, grazing should not be delayed where weevils are present.

Pearl millet *(Pennisetum glaucum)* is a very rapid growing, drought-tolerant species which is palatable and nutritious. It may be grazed or made into silage. Its growth is so rapid that, when grazed, it may require mowing to prevent growth progressing more rapidly than grazing. Sugarcane *(Saccharum officinarum)* is a high-yielding plant which, like pearl millet, is used for forage in regions with warm growing conditions throughout the year. Sugarcane requires rich land or heavy fertilizer applications. It will give excellent milk production and weight gains when fed to cattle.

Small-Grain Crops

The members of this group which are used for forage include oats, barley, wheat, and rye, with oats being the most commonly used. Oats are also the most widely adapted to different soils and climates, being used as a winter annual in Mexico, as a summer annual in Canada, and in various ways throughout the United States. The crop is hardy, fast growing, and palatable, responds well to fertilizer applications (particularly nitrogen), and is attacked by relatively few pests and diseases.

Although winter and spring ryes rank second in popularity among the small-grain crops used as annual forage, they have, in fact, some advantages over oats. Rye will germinate on drier soil and grow on wetter soils. It produces earlier herbage than oats, though it is not as palatable. Barley is more drought resistant than oats and is used in semiarid areas.

Annual Grasses

These plants are used for forage in both warm and temperate climates. Sudangrass and its hybrids with sorghum are popular temporary summer grazing crops in warm regions and have the same area of adaptation as do millet and sorgo. Sudangrass and pearl millet give similar yields in the northern Great Plains of the United States. However, millet is inferior in the central and southern Great Plains area, except in wet sites. Also, millets are sometimes preferred in the humid south of the United States because of disease problems. Sorgo will outyield both sudangrass and millet in the humid south, but is difficult to cure for hay. Yields of sudangrass hay are high; the levels reported from Maryland, Texas, and Kansas are 3.4, 3.0, and 4.0 T/ha (1.5, 1.35, and 1.8 tons/acre), respectively. Sudangrass is grown in a mixture with soybeans or cowpeas in the eastern United States, since it gives better yields and higher-quality forage than when grown alone.

Sudangrass *(Sorghum sudanense)* is a tall annual with many erect stems, narrow leaves, and a panicle which is open when mature. The spikelets are sessile. As the name suggests, this species originated in the Sudan. It was introduced into the United States in 1909 and was soon recognized as a valuable fodder plant. It is easy to establish, gives a high seed yield, and, for an annual, recovers well from grazing or cutting. This species is used for grazing and green feed. It is not normally used for ensilage since better silage can be prepared from the fodder sorghums. Although the plant is usually grown in a pure stand, sudangrass and alfalfa are sometimes grown together. Sudangrass has a good crude protein content, about 12% in the headed plant and as high as 16% in the vegetative stage.

Sudangrass (and sometimes sorghum) often depresses the yield of the subsequent crop. It has a very extensive root system, which results in the accumulation of a substantial amount of carbohydrate material in the soil. This provides an energy source for microorganisms. As these organisms increase in number, they compete with the crop for nitrogen. The resulting temporary nitrogen shortage may be overcome by using nitrogen fertilizers. The sudangrass will also deplete available soil moisture.

Sudangrass is highly nutritious for a long period of time. It tillers freely and continually sends up new shoots as the older ones mature. For hay, the best balance between yield and quality is obtained where the plant is harvested in full head or full bloom. The percentage of crude protein will then range from 7% to 9% and the crude fiber percentage will be about 33%. The nutritive value of sudangrass hay ranks below alfalfa and other legume hays. The danger to livestock from hydrocyanic acid in the young growth of sorghum is discussed in Chapter 10.

In recent years, sorghum–sudangrass hybrids have been widely used for dairy cow pastures in the humid south of the United States. These plants are crosses between sorghum *(Sorghum sudanense)* and either *S. arundinaceum* or, more frequently, *S. bicolor*. Like sudangrass, the hybrids have good feed quality characteristics. However, they have better drought tolerance, mature earlier, and give much higher yields of forage.

Sorghum bicolor is the sorgo, or syrup, sorghum. The hybrid, using this species as one parent, and *S. arundinaceum* or *S. sudanese* as the other, has sweet and juicy stems. Studies in Louisiana and Alabama have shown that grazing sorghum–sudangrass hybrids with beef steers is not advisable. In the early stages of growth, the sorghum–sudangrass hybrids may have a higher prussic acid content than sudangrass. Such material should not be grazed before it is 60 cm (24 in.) high. Climatic conditions (such as drought or frost) which cause plant stress tend to increase the prussic acid content. Where this occurs, the dangers of poisoning may be offset by allowing the plants to mature, letting frosted plants dry, or storing the material as silage.

The hybrid plants have stems and leaves intermediate in size between those of their parents. The best-known cultivars are *Sweet Sudan Grass* and *Sudax* from Texas and *Lahoma* from Oklahoma. Many commercial hybrids are also available. There is also available a short-lived perennial hybrid, which is a cross between *S. sudanense* and *S. halepense*. This hybrid is not widely used.

Annual ryegrass is used for winter pastures and lawns in warm climates and for summer pastures in temperature areas. This species was discussed in Chapter 3.

Legumes

Of the species considered in Chapter 4, arrowleaf clover, crimson clover, the vetches, the sweetclovers, subterranean clovers, and some lespedeza species are either annual or may be used as annuals. Other leguminous plants, most of which are commonly used as grain crops, may also be grown for annual forage. These are peas, soybeans, cowpeas, velvet beans, mungbeans, and peanuts. Most of these species grow well only in warm climates during the summer. Soybeans are sometimes used, and peas are used extensively as summer crops in temperate regions. Velvet beans, mungbeans, and cowpeas are tropical plants coming from India, southern Asia, and Africa, respectively. All these are widely adapted to the southern United States, give rapid initial growth, and quickly provide good soil cover. Velvet beans give very heavy yields and they are frequently grown in a mixture with a crop, such as corn, which will support the velvet bean vines. The mungbean, grown mainly in southwest Oklahoma and Texas, is susceptible to nematodes and is not suitable for the southern coastal plain of the United States. Soybeans will grow best on loam to sandy soils, but are widely adapted. For hay, this plant is cut when in mid to full bloom. If the beans have formed, the stem will be woody and many of the leaves will be lost. Soybean hay cures rather slowly.

Succulent Fodders

These plants are divided into two groups: *root* crops and *leaf and stem* crops. The root crops are not popular, since they contain a large amount of water (70% to 90%) and are consequently expensive to handle and grow. Throughout the world, the area sown to root crops such as turnips, swedes and mangolds has declined substantially in the last 50 years so that they are now rarely grown. The root crops, which have little fat, protein, or minerals, are sources of energy containing large amounts of sugar and starch. The fiber, which is found mainly in the outer skin, is cellulose and hemicellulose and is highly digestible (Chapter 9).

TABLE 5-7
Relative dry matter yields of grass and succulent forages in North America (3-year averages)

	Dry matter	
	kg/ha	lb/acre
Newly established brome–alfalfa mixture	13,000	11,570
Well-fertilized old pasture	14,000	12,460
Marrow-stem kale	17,000	15,130
Turnips and swede	12,000	10,680
Fodder beet	14,000	12,460
Corn	18,000	16,020

(Source: Various Agriculture Canada publications)

In contrast, the leaf and stem succulents like kale, cabbage, mustard, rapeseed, and buckwheat are high in vitamins (especially carotene) and protein, but frequently contain indigestible lignified fiber. They are usually of lower carbohydrate value than the root crops. Yields of leaf and stem succulents are frequently very high. Crops like marrow-stem kale have done well in Ontario and Alberta (Table 5-7). Throughout the southern United States, such crops are important to small farmers who cannot put up other stored feeds.

The disadvantage of the leaf and stem succulents is that the period of grazing is very limited. As well, storage is usually unsatisfactory because many molds develop.

FURTHER READING

Barnard, C. ed., 1964, Grasses and Grasslands, Macmillan, New York, p. 269.

Bowring, J. D., 1964, J. Nat. Inst. Agric. Bot. **10**:61.

Donald, C. M., 1963, Adv. Agron. **15**:1.

Gill, N. T., and K. C. Vear, 1969, Agricultural Botany, Gerald Duckworth & Co., London, p. 637.

Lupton, F. G. H., 1961, Ann.Appl.Biol.**49**:557.

Milthrope, F. L., and J. D. Ivins, eds., 1966, The Growth of Cereals and Grasses, Butterworth, London, p. 358.

Sprague, H. B., ed., 1974, Grasslands of the United States, Iowa State University Press, Ames, Iowa, p. 219.

Walton, P. D., 1975, Annual forages seeding rates and mixtures for central Alberta, Can. J. Plant Sci. **55**:987–993.

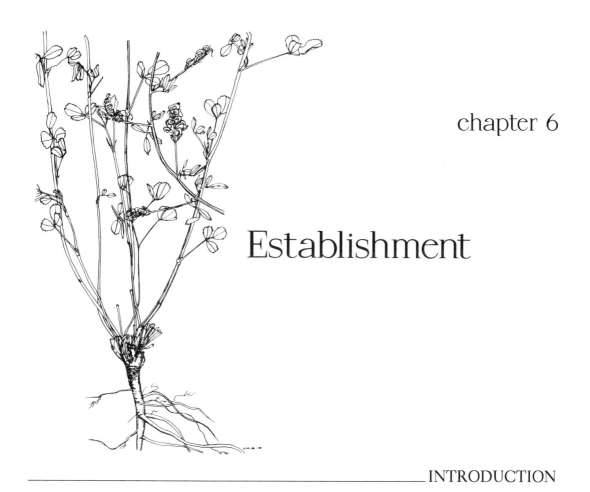

chapter 6

Establishment

INTRODUCTION

The saying "well-sown, half-grown" is usually applied to crops grown on arable land, but it is equally true for perennial pasture plants. In fact, the small-seeded perennial forage species call for more care in seedbed preparation and cultivation than do the larger-seeded row crops. Without a firm, fine seedbed, the contact area between the small forage seed and the moist soil is reduced, and germination may not take place. Where it does, many of the small seedlings will perish, because the large air spaces in a "loose" seedbed will lead to rapid moisture loss and death of the sensitive young roots.

Equal in importance with good seedbed preparation is the *quality* of the seed. There are two aspects of seed quality: high germination and genetic purity. Germination is important because in many forage species it is inherently low and quite variable. A guarantee of genetic purity is especially vital, because, while it is possible to identify the seeds of different forage species, it is quite *impossible* to distinguish one forage cultivar from another by examining the seed alone. Genetic purity is difficult to maintain since most forage species are cross-pollinated. Most countries have government supported seed-certification agencies which maintain cultivar pedigrees and issue *labels* to indicate seed-quality classes. The Canadian Seed Growers Association is an

TABLE 6-1
Relative moisture requirement and soil tolerance for forage species

	MOISTURE REQUIREMENT						TOLERANCE				
	Good moisture and poor drainage	Good moisture and drainage	Semi-moist	Semi-dry	Dry	Very dry	Partial flooding	Severe flooding	Moderate salinity	Severe salinity	Moderate acidity
LEGUMES											
Alfalfa		X	X	X	X	X					
Alsike clover	X	X	X				X				X
Red clover		X	X								X
Sweet clover		X	X	X					X		
Birdsfoot trefoil	X	X	X	X							X
GRASSES											
Bromegrass		X	X	X							X
Creeping red fescue	X	X	X								X
Orchardgrass		X	X								
Reed canarygrass	X	X					X	X			X
Russian wildrye				X	X	X			X		
Tall fescue	X	X	X				X		X	X	X
Timothy	X	X	X				X				X
Crested wheatgrass			X	X	X	X			X		
Pubescent wheatgrass		X	X	X	X				X		
Intermediate wheatgrass		X	X	X	X						
Tall wheatgrass	X	X							X	X	
Slender wheatgrass		X	X	X					X		

agency of this type. The quality of seed which does not carry a valid certificate of guarantee is always doubtful. This topic will be discussed further in Chapter 15.

In the following pages, we will consider the problems which confront a farmer who has decided to establish a new perennial pasture. We will examine the problems faced in choosing species and cultivars and study the use and selection of a forage mixture. The seed quality, germination, the seedbed, planting depth, seeding rate, time of seeding, soil fertility, and the use of companion crops will also receive attention.

CHOICE OF CROPS

For almost all locations, provincial, state, or federal agencies have available yield data for locally adapted species and cultivars. This information should be consulted when choosing a forage crop, since local knowledge is most important. General recommendations for temperate forage species are set out in Table 6-1.

SELECTION OF MIXTURES

Where it has been possible to select a forage that is *very well* adapted to a given location, growing any other species with it in a mixture is likely to result in lowering the total herbage yield (Table 6-2). Why then are most forage crops grown in mixtures? The answer to this involves two major considerations, as well as a number of valid minor ones. The two main concerns both arise from the fact that total annual yield is not the most important feature of forage production. First, animals require a consistent level of nutrition throughout the year. Forage species reach their peak production at different times during the growing season (see Chapter 17 and Figure 17-4). Uniform production will, therefore, most likely be obtained by using a mixture of species.

Second, *quality* of feed is important for high animal productivity and, if grasses alone were used, available protein could well limit animal production. Legumes, which form a symbiotic relationship with bacteria of the genus *Rhizobium* and so fix atmospheric nitrogen (see Chapter 1), increase the average protein content of the herbage over that obtained from a grass by itself. Also, growing a legume with a grass increases total herbage yield beyond that of a legume grown alone so that animal intake and

TABLE 6-2
Yields of orchardgrass and reed canarygrass grown in monoculture and in mixture in Elk River, Minnesota

	Yield	
	T/ha	tons/acre
Orchardgrass	7.71	3.5
Reed canarygrass	7.33	3.3
2 Orchardgrass: 2 reed canarygrass	7.61	3.4
2 Orchardgrass: 1 reed canarygrass	7.59	3.4
1 Orchardgrass: 2 reed canarygrass	7.29	3.3
1 Orchardgrass: 1 reed canarygrass	7.61	3.4

(Source: Sheaffer et al., 1981, Agron. J. **73**. By permission of The American Society of Agronomy, Inc.)

FIGURE 6-1
A mixture of creeping red fescue, smooth bromegrass, and alfalfa gives a complete ground cover of nutritious herbage.

weight gains are increased. These two main features are well illustrated by the pasture mixture in most common use in Alberta (smooth bromegrass, creeping red fescue, and alfalfa). Smooth bromegrass gives high yields over the major part of the growing season, while creeping red fescue provides green, high-protein herbage in the fall, thus helping to give consistent production. Alfalfa increases the animal intake and the protein content of the mixture (Figure 6-1).

Additional reasons for using a mixture arise from the vast variation that exists in the macro- and microenvironments provided by both climate and soil. Fertility level, pH, and soil moisture will vary over a field. Temperature, light intensity, and moisture will vary within and between years. The performance of the different forage species will *not* rank in the same order for all these diverse environmental circumstances. Thus, seeding and cropping hazards will not be the same for all species. Indeed, it is difficult to find one species that will yield well under all circumstances. Only a mixture will provide uniform production within and between growing seasons and locations.

Species of different height and with root systems which utilize different soil strata will make the best use of light, soil nutrients, and moisture; thus, production will be high and weed invasion less likely. Complete soil cover will also reduce erosion risks and hold the soil steady, thereby preventing legumes from being pushed out of the soil should it move when freezing. A mixture will also reduce the risk of insect and disease attack. Some such organisms will infest a wide range of different species (e.g., grasshoppers), but many are species specific and a mixture will prevent the buildup of a large infestation.

None of this should be taken to indicate that mixtures should contain a large number of different species. Indeed, the most widely used mixtures are simple and often contain a single grass and a single legume. Smooth bromegrass and alfalfa are used in dry areas; timothy and red or alsike clover grow where soil moisture is more plentiful. In the warm, humid south of the United States, bermudagrass is grown with lespedeza or white clover, and soybeans are grown with millet, sudangrass, or sorghum.

MAKING A FORAGE MIXTURE

In making up a forage mixture, a number of important objectives must be met. The species included in the mixture should be appropriate to the purpose for which the pasture is grown. Tall grasses and legumes are suitable for hay production. Lower-growing types are used for grazing. If the pasture is to be cut for hay, it is important that each of the species in the mixture reaches the *hay stage* (the optimum balance between quantity and quality, see Figure 9-6) at the same time. On the other hand, if the pasture is mainly for grazing, such uniformity is not important and diversity may be advantageous.

It is desirable that the components of a mixture be compatible. Species with similar palatability, aggressiveness, and adaptation will result in a stable mixture. In general, rhizomatous and bunch grasses are *not* compatible. However, a bunch grass may grow well with a creeping legume (e.g., orchardgrass and white clover), and spreading grasses are compatible with noncreeping legumes (e.g., smooth bromegrass and alfalfa). Species similarity in palatability is important for a pasture that will be grazed, but this feature is much less important if the crop is to be used for hay.

Forage mixtures should also be designed with the object of reducing the risk of digestive disorders in the animals that will graze it. The risk of bloat is a factor which frequently determines the proportion of alfalfa in a seed mixture (see Chapter 10).

SEED QUALITY

First, let us consider the reasons why varietal purity is important. The forage breeder carefully selects parent plants which not only give good yields of high-quality herbage, but also cross together to produce progeny that perform well. Forage species are primarily cross-pollinated. Hence, their progeny is heterozygous and will segregate, in subsequent generations, to produce plants which are likely to be poorer yielding or of poorer quality than the parental types originally selected by the breeder. Furthermore, where such progeny are cross-pollinated with unselected plants (rather than with each other), deterioration will be even more extensive.

Only seed which carries a pedigree certificate from a national or state agency or a similar guarantee from the producer can be regarded as having been produced under circumstances which maintain a satisfactory genetic background. Also, by purchasing seed that has been produced in a pedigree seed multiplication scheme, the farmer obtains seed in which weeds, disease, and insect populations are low. The seed crops in the field and the seed are both inspected to make certain that they are relatively weed free. However, some American states are now certifying seed for genetic purity only. No attempt is made to evaluate germination and mechanical purity. Well-managed seed multiplication schemes (see Chapter 15) are implemented in areas where diseases and insects are not prevalent. The use of a disease- and insect-free area is especially important for grass seed production.

GERMINATION

A high germination rate is an important aspect of good seed quality. The topic is treated separately here because it has a vast literature, having received extensive attention from botanists. During the life of almost all the higher plants, there is a quiescent stage of reduced water content. This period enables the plant to survive unfavorable conditions and permits dispersal. For the Spermatophyta, this part of the life cycle is expressed by the formation of seed and terminated when the young sporophyte is active again. This statement could, broadly speaking, define germination. However, a series of definitions are required to accommodate all the ways in which the word "germination" is used. To study these definitions, let us examine the germination process.

The prerequisites for germination are a viable seed, adequate moisture, and a favorable temperature regime. The absorption of water is very rapid, with 80% of the saturation amount being taken up during the first few hours of immersion. The maximum rate of absorption occurs during the first hour. As free water becomes available in the seed, stored substances move into solution. The earliest step in germination is the initiation of the activities of existing cells in the embryo. In grasses, cell extension starts in the coleorhiza, through which the primary root emerges. "Germination," when used as a scientific term, is usually taken to mean the appearance of the radicle outside the seed coat. This is the definition used in laboratory germination tests; "85% germination" indicates that 85% of the seeds produced radicles. The farmer is concerned with developments at the other end of the embryonic axis and, for agricultural circumstances, "germination" means the emergence of the plumule above the soil surface. On the other hand, for legal purposes, a seed is only deemed to have germinated if it has been shown to be capable of producing a plant.

As morphological changes take place, chemical events in the seed result in a sharp rise in respiration rate. This change is associated with the development of the embryo, rather than the degradation of the starchy endosperm. The embryo directs the activation or synthesis of the large number of enzymes which catalyze the reactions leading to growth and development. The embryo produces and secretes into the endosperm enzymes causing the release of proteins into solution.

Evolution has played an important part in the development of germination patterns. A lack of uniformity in the time of germination is of evolutionary advantage if initial germination encounters unfavorable growing conditions. In the grasses, natural selection has created considerable diversity for time of germination. In contrast, from an agricultural point of view, uniform germination is an advantage. Such selection has been applied by people, both consciously and unconsciously, to our annual cereal crops over the thousands of years that they have been cultivated. Most forage crops have been in agricultural use for a much shorter time than the cereals, and, in any case, selection pressures would not bear so heavily on a perennial species. As a result, laboratory germination tests for forage grasses usually give lower values than those for the closely related cereal crops. These low germination percentages often indicate not a lack of *ability* to germinate, but rather that germination did not take place during the time period (often 10 to 15 days) over which the test was conducted.

Let us consider now the kinds of delaying mechanisms which evolution has built into the germination of the grasses. The grass embryo is capable of continued development soon after pollination. There will be a period of 1 or 2 days, or weeks, in which up to 25% of the newly formed embryos are capable of development. Germinative ability will then decline, and at the time when the seed is morphologically mature, or ripe, germination *cannot* take place. In many grasses, this stage is very brief (e.g., Italian ryegrass, timothy, and meadow fescue). Other species (e.g., *Paspalum*) are said to exhibit *dormancy* or *primary dormancy* and will not germinate for weeks or months. Thereafter, germination will be influenced to varying degrees by temperature and light, depending on the species. The behavior of members of the same genus is markedly divergent. For example, tall and creeping red fescue are not influenced by light, while the germination of meadow fescue is greatly promoted by illumination. However, light sensitivity is found in the members of most tribes of the Festucoideae, Panicoideae, and Chloridoideae. This divergence of species requirements means that detailed knowledge of favorable conditions for germinating one species may not be helpful in overcoming difficulties encountered with another species. With this in mind, we will consider some procedures that frequently enhance the germination of the seeds of species used for forage production.

Temperature. Where a single constant temperature is used, the optimum for most grasses is between 15° and 25°C. Alternating temperatures will enhance germination. The higher temperature is normally restricted to 6 to 8 hr out of the 24-hr cycle. Since the sharpness of alternation is the effective feature, changes should be rapid and move through not less than 10°C. Redtop and orchardgrass require temperature alternation for germination. Prechilling and holding imbibed seeds at 4° to 10°C (39° to 50°F) for a few days, followed by a transfer to optimum temperature conditions, frequently enhances germination.

Light Intensity. Effective intensities are usually within the range from 5 to 100 fc. Some grasses are entirely indifferent to light, while the germination of many nongramineous species is controlled by photoperiod.

Chemical Treatments. A large number of compounds will influence germination. The most commonly used substances are potassium nitrate (0.1% to 0.2%) and thiourea (0.1% to 1.0%). In recent years, the use of gibberellic acid to "break dormancy" or hasten germination has received much attention. In the gramineous species, it seems that gibberellic acid will act, in effect, as a "substitute" for light, where such a requirement exists. This is not the case for the nongramineous species.

Modification of the Seed Coat. The methods of improving germination discussed so far are research techniques which are not normally used by the farmer. In contrast, seed-coat modification is a widely used agricultural practice. For both grasses and legumes, the object is to cut, wound, or break the seed coat in such a way as to remove, either by hand or by milling, the covering structures. This process is called *scarification*. In the legumes, it is designed to scratch the seed coat of "hard" seeds and allow water to penetrate. Few grasses have impermeable seed coats (buffalograss being one), but the covering structures may prevent germination by mechanically restraining the radicle. Also, the removal of the outer parts of the grass seed facilitates gaseous exchange, water absorption, and either the actual removal of the germination inhibitor substances found in many species or their leaching from the remaining seed coat, the endosperm, or the embryo.

Harvesting and Storage. Seed maturity at the time of harvesting has a significant effect on germination, and insofar as harvesting method is associated with maturity, it too will effect germination. For example, Russian wildrye heads shatter readily, and the crop must be cut before it is fully ripe when the seed is in the "firm dough" stage. Immaturity may be avoided and germination improved (Figure 6-2) by swathing the crop just above the leafy basal growth. The heads are then threshed with a combine several days later.

For most forage species, a storage temperature of 10°C (50°F) with a relative humidity of 50% or less is satisfactory. There is, however, considerable variation between species. Many grass species, after reaching physiological maturity, become dormant again, a condition called *secondary dormancy*. For all species, there is a reduction in germinative ability as the seed ages. The seedlings produced from such seed frequently grow slowly. The nature of this process is not well understood.

SEEDLING VIGOR

Following germination, the seedling passes through a *heterotrophic* stage, during which time it obtains complex organic compounds from the endosperm, to the *autotrophic stage*, in which it depends on its own photosynthetic products. There is frequently an intermediate stage in which nutrients are derived from both sources. These metabolic processes coincide with establishment, a time of intense competition between members of the same and different species. Plants which are able to survive, especially if environmental conditions are unfavorable, are said to have *seedling vigor*.

No single attribute provides an adequate description of seedling vigor. The ability to emerge, to withstand dry or cold conditions, to resist the attack of soil- and seed-borne microorganisms, as well as the ability to compete with other plants, have all been regarded as expressions of seedling vigor. In the widest sense, seedling vigor must involve an increase in size of any one plant which is more rapid than other plants of the same age. Prompt growth under favorable conditions is an obvious part of successful establishment. When one or more of the factors such as air, light, nutrients, and water are limited or subject to depletion, the plant which can exploit the greatest volume of

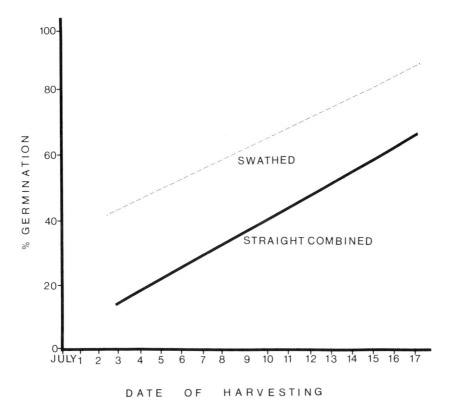

FIGURE 6-2
Germination of Russian wildrye when harvested by two methods. (Source: T. Lawrence, 1977, Agriculture Canada Publication 1607)

the environment will be the successful competitor. What are the characteristics of a plant of this type?

First, seed size and weight are associated with seedling vigor. The seeds of most forage crops are small, but even among the seeds of one plant, there is considerable variation, so that the largest seed may weigh ten or more times as much as the smallest. Germination percentages and seedling growth rates are always lower in small and immature seeds. This difference is determined by the relative volume of endosperm.

Second, a high level of biochemical and physiological activity is essential to seedling vigor. Only the rapid mobilization of reserve material in the endosperm and the transport of nutrient substances to growing points will sustain cell division and elongation. For example, reed canarygrass will use all the nutrient material from the endosperm within 10 to 14 days of germination. A number of rapidly growing weeds, which successfully compete with reed canarygrass seedlings, will mobilize all endosperm reserves in 24 to 48 hr.

Closely associated with such metabolic events is the force of seedling growth, that is, the ability of the plant to grow through the soil. This character is dependent on the strength and durability of the epidermis, as well as the extent of the endosperm reserves. Lawrence, working at Swift Current, Saskatchewan, showed that seed and forage yield were correlated with the force of seedling growth. Rapidity of germination and the growth rates of both the root and shoot system are also reflections of the seedling's metabolic activity. The development of the shoot is important so that a large photosynthetic area may be established early in the life of the seedling. A large root system is required to gather water and soil nutrients. The size of the endosperm is especially important in relation to the development of the root system. Most of the early photosynthetic products are used for shoot growth, and the endosperm must provide the major part of the nutrients for early root extension.

Genetic selection for seedling vigor could contribute and, in limited cases, has contributed much to the improvement of forage cultivars. Selecting for seed size is especially promising, since the heritability for this character is frequently high. The genetics of many of the various biochemical and physiological processes involved in seedling vigor still remain to be determined.

SEEDBED AND SEEDING DEPTH

The ideal seedbed for forage crops is one in which a loose layer of very fine soil to cover the seed overlies compacted soil. The small forage seeds require, for optimum emergence, a planting depth between 10 and 65 mm (0.5 and 2.5 in.) (Table 6-3). Within this range, the depth for individual species varies with their seed size. It is evident that forage seeding calls for very precise control of seeding depth. This is only possible under uniformly firm soil conditions. Hence, a firm seedbed is desirable both because it retains moisture and because it permits precise depth control. Such a seedbed may be attained by a fall plowing to a depth of about 12 cm (about 5 in.). In areas where there is a danger of erosion by wind or water, seedbeds are left somewhat "lumpy." Apart from the influence on soil structure, plowing is important for weed control. There is nothing to be gained from a deep plowing, since the forage plant has a small root system in the first year.

One way of protecting against erosion and weeds is to use a lightly seeded companion crop of oats. Alternatively, the forage crop may be planted in the stubble of a preceding cereal crop. In this case, plowing is usually omitted and a light discing or harrowing is the only cultivation required. Where the land is plowed, it is essential that land preparation be started early so that the soil will settle and be "firm" before it is seeded. Heavy rain will do more to compact the soil than many roller or cultipacker treatments and will most certainly cost less!

Some soils will form an impenetrable crust when a fine compacted seedbed is prepared. Under these circumstances, seedling emergence is restricted to cracks in the soil surface, and plant populations are substantially reduced. In such cases, the number of cultivations must be reduced, or the possibility of establishment without mechanical cultivation should be considered.

TABLE 6-3
Effect of depth of seeding on percentage emergence

	Depths of seeding, mm (in. in parentheses)				
	13 (0.5)	25 (1)	38 (1.5)	50 (2)	76 (3)
Alfalfa in sand	71	73	55	40	—
Alfalfa in loam	60	55	32	16	—
Alfalfa in clay	52	48	8	13	—
Clover, alsike	53	49	9	4	—
Clover, red	56	62	22	14	—
Clover, ladino	47	28	2	0	—
Bromegrass in sand	70	64	48	29	—
Bromegrass in loam	68	50	31	19	—
Bromegrass in clay	56	37	17	5	—
Timothy	89	81	39	12	—
Red top	64	33	2	0	—
Crested wheatgrass	87	—	44	—	0
Red canarygrass	76	—	67	—	9
Russian wildrye	93	78	80	15	0
Slender wheatgrass	99	—	97	—	41
Tall wheatgrass	93	—	83	—	3
Kentucky bluegrass	43	27	4	1	—
Sweetclover	51	45	26	14	—

(Source: Collected from various publications by Agriculture Canada and Alberta Agriculture)

The main value of mechanical cultivation lies in weed control, rather than in the loose soil structure it develops. If weeds can be killed by using herbicides, then the minimal mechanical cultivation produced by the seed drill when the crop is sown will provide all the tillage required. The use of herbicides will depend on the botanical composition of the existing herbage and on the species to be established. Attempts to develop such technology started in the latter part of the last century and have met with the greatest success in the renovation and improvement of existing grasslands. This frequently involves the establishment of a legume species in an existing grass sward. The old pasture should first be weakened by repeated clipping or grazing for several weeks prior to seeding. A herbicide is then used to reduce grass competition with the legume seedlings. Frequently, the herbicide is applied only in the rows in which the legume is to be seeded. Before or at the time of seeding, lime and fertilizer are applied to meet the needs of the legume, and are placed in a band below the seed. The amount applied should be based on information from a soil test. Satisfactory coverage of the seed is vital. Together with unfavorable weather after seeding, poor seed coverage is the most common cause of failure for this type of establishment.

The use of herbicides in forage establishment is not widespread. Apart from concerns about environmental contamination, the cost of herbicides as compared to mechanical cultivation and the limited availability of satisfactory chemicals are the reasons for the lack of popularity of this method of establishment.

Under the harsh environmental conditions which are common on the rangelands of the Great Plains and prairies of North America, sufficient moisture may not be available for seedling emergence when the soil surface is bare. Artificial and natural mulches may then be used to modify temperature and moisture. While stubble from a previous crop is the most common type of mulch, weathered hay or straw applied in a 30 cm (about 10 in.) wide strip over the rows at a rate of 2,000 kg/ha (1,780 lb/acre) has been shown to reduce soil temperature and increase available soil moisture. A variety of artificial mulches, often made of polyethylene, are also available, but these are generally regarded as being too expensive for forage establishment under rangeland conditions.

SEEDING EQUIPMENT

There are three ways in which forage seeds are sown. First, they may be broadcast by hand or from a wind seeder. Second, a grain drill with a grass seeding attachment may be used. Third, corrugated rollers with grass seed attachments, usually called a "cultipacker," are available. With this last type of equipment, the forage seed can be sown with different species being seeded at different rates from separate seed boxes, a companion cereal crop can be established, and fertilizer can be applied all at once. The fertilizer is placed in a band at a depth of about 40 mm (about 1.5 in.), and the forages are seeded into the soil above it. In many cases, a depth-of-seeding device, consisting of a distance ring on the inside of the seed disc cutting edge, prevents the seed from being placed too deeply in the soil. Such devices not only eliminate seeding depth as a major cause of stand failure, but also reduce the need for extensive seedbed preparation. The disc will be maintained at a constant depth over a seedbed even if the degree of compaction varies.

TIME OF PLANTING

When determining the time of planting, there are two prerequisites for forage crop establishment. The first is to plant at a time when the moisture at the soil surface is adequate for germination. The second is to allow as long a growing season as possible so that the new crop will consist of large and vigorous plants that can withstand the winter, when temperatures will be unfavorable and soil moisture limited. Throughout much of the North American continent, these two prerequisites are most likely to be met by seeding in spring or early summer. In the southern United States, these conditions are often met by planting cool-season species in the fall. At that time, rainfall is good and the potential length of the growing season is at a maximum. If the land is weedy, a late summer seeding will allow time for cultivation. Midsummer seedings are to be avoided, since high temperatures will dry out the soil, and seedling diseases are most common during this time.

The ideal spring seeding time for forages is also the ideal time for seeding many other crops. Where the seeding times for forages and crops which bring an immediate cash return coincide, the choice usually favors the cash crop. Spring is, in any case, a busy time on any farm, and we should consider alternative times for forage crop establishment. In areas where there is a marked winter season, it is possible to seed just before freeze-up. The outcome is much the same as a very early spring seeding, for the seed will germinate in the following spring. There are some species whose seed will not overwinter in the field (e.g., sweetclover), while many other species will only partially survive the fluctuating temperatures in the spring.

Young seedlings, up to the fourth-leaf stage, are very sensitive to low temperatures and are quite unable to *harden* (i.e., change physiologically into a state which will enable them to withstand low temperatures). Early fall plantings, which permit the young plants to reach the fourth-leaf stage before the onset of winter, are frequently successful in giving a full forage crop in the following year. Such early planting is especially important in the southern United States, because weeds are less likely to cause problems at that time.

SEED RATES

Many local circumstances determine the amount of seed which should be used to establish a new stand. The amount and distribution of rainfall, the physical and chemical properties of the soil, and the nature of the seedbed are all important environmental considerations. The germinative ability of the seed, the nature of the species or varieties to be used, and the purpose for which the forage is to be established are seed-related management considerations. Together with recommendations on the local adaptation of species and cultivars, many provincial and state agencies provide information on seed rates for suggested forage mixtures. Such recommendations should certainly be carefully considered. As a rough guide, the total amount of seed required for seeding most mixtures is about 10 kg/ha (9 lb/acre). This makes allowance for a substantial seedbed loss due to poor germination, lack of aeration, and unfavorable temperature or moisture conditions. Given ideal establishment conditions, 2 kg/ha (1.75 lb/acre) of seed of many forage species should be adequate for satisfactory establishment (see Table 6-4), but such circumstances are not frequently encountered.

SEEDING PATTERN

There is now a considerable amount of information available which indicates that, for most forage crops, row planting gives superior stands and yields to broadcast seeding. There is also an ever-increasing body of information to indicate that the seeding of grasses and legumes in alternate rows gives striking yield advantages. Such methods have shown annual yield increases of about 25%. These increases are greatest in years

TABLE 6-4A
Seed characteristics for legumes and grasses (metric units)

Species	Average number of seeds per kg	Plants per m² seeded at 1 kg/ha
Alfalfa	484,000	48.4
Bahiagrass	330,000	33.0
Bentgrass		
Colonial	18,700,000	187
Creeping bent	17,160,000	1,720
Velvet bent	24,200,000	2,420
Bermudagrass	3,960,000	396
Birdsfoot trefoil	713,000–1,200,000	71–120
Bluegrass		
Canada	5,500,000	550
Kentucky	4,840,000	484
Bromegrass, smooth	297,000	29.7
Cicer milkvetch	286,000	28.6
Clover		
Alsike	1,540,000	154
Red	550,000	55
White	1,760,000	176
Dallisgrass	485,000	48.5
Fescue		
Creeping red	1,353,000	135
Meadow	506,000	50.6
Tall	506,000	50.6
Johnsongrass	290,000	29.0
Lespedeza	820,000	82.0
Lovegrass, weeping	3,300,000	330.0

when environmental conditions are least favorable. For example, in southern Saskatchewan, during a dry year, the alfalfa component of a mixture, not in alternating rows, contributed 175 kg/ha (156 lb/acre). When the rows of grass and alfalfa were alternated, the alfalfa yield was 650 kg/ha (579 lb/acre). Alternating rows reduces the competition between the legumes and the grass during establishment. A corresponding increase may be obtained by "cross" seeding the grass and legume.

SEED TREATMENT

Attention has been drawn earlier in this chapter to the improvements in germination that may be produced by scarification. There are other forms of seed treatment which will aid establishment and improve yield.

TABLE 6-4A (Continued)

Species	Average number of seeds per kg	Plants per m² seeded at 1 kg/ha
Oats	30,000	30.0
Orchardgrass	1,430,000	143
Redtop	11,000,000	1,100
Reed canarygrass	1,166,000	116.6
Rhodesgrass	4,700,000	470
Ryegrass		
Altai wild	581,000	58.1
Annual (Italian)	500,000	50.0
Perennial	506,000	50.6
Russian wild	385,000	38.5
Sainfoin	50,600	5.0
Sudangrass	121,000	12.0
Sweetclover	572,000	57.2
Timothy	2,700,000	270
Vetch		
Crown	304,000	30.4
Hairy	46,000	4.6
Wheatgrass		
Crested 2n	704,000	70.4
4n	418,000	41.8
Intermediate	194,000	19.4
Pubescent	242,000	24.2
Slender	352,000	35.2
Streambank	343,000	34.3
Tall	174,000	17.4

(Source: Compiled from several sources, including Agriculture Canada Publication 1405)

Inoculation. To provide an adequate number of bacteria of the genus *Rhizobium* for the successful development of a symbiotic relationship with most legume crops, inoculation is required. The genus *Rhizobium* consists of gram-negative aerobic bacteria which may exist as a *coccus* or a *bacillus*. The coccus is flagellate and capable of moving in the soil. They are nonsporulating and aerobic, with an optimum growth temperature of 25°C (77°F). These organisms may be divided into a number of inoculation groups which, while identical in appearance, are capable of forming a symbiotic relationship with only one type of legume. The number of inoculation groups listed in the literature varies. However, a different inoculum is required for the following species or groups: alfalfa; the true clovers; pea and vetch; bean; soybean; lupine; and cowpea. Within each of the inoculation groups, there is a wide range of physiological

TABLE 6-4B
Seed characteristics for legumes and grasses (English units)

Species	Average number of seeds per lb	Plants per ft² seeded at 1 lb/acre
Alfalfa	220,000	4.9
Bahiagrass	150,000	3.3
Bentgrass		
Colonial	8,500,000	19
Creeping bent	7,800,000	172
Velvet bent	11,000,000	242
Bermudagrass	1,800,000	40
Birdsfoot trefoil	324,090–545,000	7–12
Bluegrass		
Canada	2,500,000	55
Kentucky	2,200,000	48
Bromegrass, smooth	135,000	30
Cicer milkvetch	130,000	29
Clover		
Alsike	700,000	15
Red	250,000	6
White	800,000	17
Dallisgrass	220,000	5
Fescue		
Creeping red	615,000	14
Meadow	230,000	5
Tall	230,000	5
Johnsongrass	138,000	3
Lespedeza	372,000	8
Lovegrass, weeping	1,500,000	33

strains which may be specific to a few cultivars. They react less favorably or not at all with other cultivars of the same species group. Under these circumstances, the selection of the most appropriate physiological strain is critical.

In many parts of North America, soils are well-stocked with bacteria which will associate with alfalfa, sweetclover, and the true clovers. As the legume plant germinates, the flagellate *cocci* in the soil assemble around a root hair, which subsequently grows around the bacterial colony. The wall of the root hair breaks down, enabling the bacteria to move into the plant and migrate toward the cortical region. The bacteria multiply to form a swelling or root *nodule* and change in form to become *bacilli*. The plant develops a vascular structure which surrounds the nodule to facilitate the movement of sugars provided by the plant and nitrate substances formed by the bacteria fixing atmospheric nitrogen.

TABLE 6-4B (Continued)

Species	Average number of seeds per lb	Plants per ft² seeded at 1 lb/acre
Oats	13,600	3
Orchardgrass	650,000	14
Redtop	5,000,000	110
Reed canarygrass	530,000	12
Rhodesgrass	2,136,000	47
Ryegrass		
Altai wild	264,000	6
Annual (Italian)	227,000	5
Perennial	230,000	5
Russian wild	175,000	4
Sainfoin	23,000	0.5
Sudangrass	55,000	1
Sweetclover	260,000	6
Timothy	1,227,000	27
Vetch		
Crown	138,000	3
Hairy	20,900	0.5
Wheatgrass		
Crested 2n	320,000	7
4n	190,000	4
Intermediate	88,000	2
Pubescent	110,000	3
Slender	160,000	4
Streambank	156,000	3
Tall	79,000	2

(Source: Compiled from several sources, including the U.S. Department of Agriculture and Agriculture Canada Publication 1405)

The plant develops nodules when it is about 2 weeks old, and it may take 1 or 2 weeks for the vascular system to develop and for the nodule to become active. Until that time, legume plants will respond well to applications of a nitrogenous fertilizer. However, this practice is not recommended, since it retards the development of fully active bacterial nodules.

The root system of the legume plant is reduced when the plant is cut or grazed (or when it dies). At these times, many nodules are shed into the soil, where they break down and liberate some nitrate substances and bacteria (in the cocci form). These cocci are then capable of forming a symbiotic relationship with the same, or another, legume plant. With such bacteria present in the soil, there *may be* no need to inoculate.

TABLE 6-5
Nitrogenase activity per plant as influenced by water regime in California

	Ethylene per plant per season, mmoles	
	Inoculated	Uninoculated
Well-watered	14.5	2.1
Drought	3.7	0.2

(Source: Zablotowicz et al., 1981, Agron. J. 73. By permission of The American Society of Agronomy, Inc.)

However, the bacteria are killed by high temperatures (60°C or 140°F), direct sunlight, low pH, dessication, and lack of oxygen (possibly due to flooding). Also, drought will reduce the number of effective nodules (Table 6-5). Therefore, it is recommended that small seed legumes should always be inoculated. This is especially true if soils are at all acid (pH 6.5 or lower). Lime requirements for soils in humid regions will vary from 1 to 5 T/ha (0.5 to 2.5 tons/acre). Peanuts, beets, alfalfa, sunflower, and sweetclover plants, as well as the *Rhizobium,* are all very sensitive to acidity.

There are circumstances in which certain strains of bacteria will form nodules which are ineffective. If cut open, these nodules usually will not show the bright red tissue color which is typical of the effective nitrogen-fixing nodule. Effective inoculation is accomplished by mixing locally adapted bacterial cultures with seed before sowing. Increased nodulation occurs if a sticker (frequently corn syrup) is used to hold the bacteria to the seed coat. The bacterial cultures are very cheap and should be in general use.

Although complicated, the biochemistry of nitrogen fixation has been extensively studied and its principles elucidated. Bacteroids, which are different from the bacteria in both form and metabolism, develop inside the host, enclosed by membranes. After the bacteroids have formed, the nitrogen-fixing enzyme nitrogenase appears in the nodule. Leghemoglobin is also associated with nitrogen fixation, but its role is not clear. Bacteroids washed free of hemoglobin can still fix nitrogen. It may be that the bacteroids require a rich supply of oxygen and that leghemoglobin provides this. Ammonia, the primary product of nitrogen fixation, is converted in the nodules to alpha amino compounds.

The elements iron, copper, cobalt, and molybdenum are important for nitrogen fixation. Copper and iron are required for hemoglobin synthesist. Molybdenum is involved in, and cobalt is essential for, nitrogen fixation. For each molecule of nitrogen fixed, 15 molecules of ATP and 6 electrons (equivalent to a total of 24 ATP molecules) are required by the nitrogenase enzyme. The plant photosynthates are translocated to the nodules and provide the electrons and the ATP. Carbon skeletons from plant photosynthates are used when ammonia, from nitrogen fixation, is converted into amides and amino acids. Aspartic acid, asparagine, glutamic acid, and glutamine are the principal products translocated from the nodule to the shoot.

It is evident that nitrogen fixation is achieved only at a certain cost to the plant. Estimates indicate that, for every milligram of nitrogen fixed, pea plants require about ten times that weight of carbohydrate to support nodule growth, respiration, and the transport of fixed nitrogen to the leaf. There is experimental evidence to indicate that the cost to the plant is the same if fertilizer nitrogen, rather than biological nitrogen, is used. The cost to the farmer is, however, substantially less!

In conclusion, the factors which influence symbiotic nitrogen fixation include the availability of plant photosynthates, plant stress factors (e.g., drought will depress nitrogen fixation, Table 6-5), shoot-sink activity, bacterial strain, plant variety, mineral nutrition, and cultural practices.

Seed Pelleting and Coating. A seed is said to be *pelleted* when it is coated with an inert substance which makes it larger and more uniform in size. Pelleting acts as a seed extender and is useful for precision planting of small forage seed. Seed *coating* can serve these functions but can also be used to enclose *Rhizobia* with the seed. For grass seed, fungicides (e.g., thiram), herbicides (e.g., dalapon), and insecticides have been used in coated seed.

COMPANION CROPS

The rationale for the use of a *companion* crop (sometimes called a *nurse* crop) is that the small forage seedling makes little demand on soil nutrients or moisture during the early months of its life. Consequently, it is possible to grow the forage crop simultaneously with an annual crop. This may well be true in the humid areas of Europe, Britain, and the United States, and for some regions of Canada, where moisture levels are high. In these areas, not only are nutrients and moisture adequate, but the season is long enough to permit forage growth to take place after the companion crop has been removed. Prior to this, the companion crop intercepts much of the incident light and uses most of the soil moisture and nutrients.

The same may *not* be said of the drier parts of Canada, Australia, and the United States. In these areas, the growing season is frequently short. Then the competition with the companion crop will be markedly detrimental to the forage crop, and since there is no opportunity for recovery after harvest, the loss to the forage crop will be substantial. Table 6-6 shows that annual ryegrass reduces wheat yield in Oregon when these two crops are grown together. However, under favorable conditions, this loss may be offset by the financial gain to be made from utilizing the companion crop. Spring-seeded cereals and flax are the most common companion crops. Flax provides the least competition for light and soil moisture and is consequently preferred. However, barley or oats frequently give a better economic return. In the southern United States, oats is prefered. Soybeans and peas have also been used successfully. Where soil drifting is a hazard, a cereal seeded at a low rate (about one-half normal) will protect the soil. A companion crop will also help to trap snow or act as a living mulch, and so provide winter protection for the forage crop. Another advantage of a companion crop is that it helps to suppress weeds which compete with the forage crop. Since weeds frequently cause more damage than a companion crop, a companion crop is to be preferred in weedy areas. In the southern United States, companion crops allow an earlier use of the pasture.

TABLE 6-6
Grain yields of four winter wheats grown at three ryegrass densities

Cultivar	Ryegrass density, plants/m^2	Wheat grain yield		Grain yield reduction
		kg/ha	lb/acre	%
Druchamp	0	3,440	2,752	0
(tall)	40	2,800	2,492	19
	99	2,480	2,207	28
Yamhill	0	4,360	3,880	0
(tall)	40	3,250	2,892	26
	107	3,010	2,679	31
Nugaines	0	3,380	3,008	0
(semidwarf)	39	2,560	2,278	24
	96	2,120	1,887	37
Hyslop	0	3,850	3,427	0
(semidwarf)	44	2,850	2,280	26
	99	2,350	2,091	39

(Source: Appleby et al., 1976, Agron. J. 68)

GRASSLAND RENOVATION

Grassland renovation is a system of conservation tillage which seeks to improve existing cultivated or native grass sods with minimum or no tillage. The advantages of retaining existing grasses are that they will minimize the risk of wind or water erosion, reduce the energy and seed requirement for reestablishment, and increase forage production in the establishment year. Improvements may be achieved by sowing legumes into an existing grass so that the two plants will grow together. For example, in the southeastern United States, winter annuals may be seeded into tall fescue pastures or, in the north, alfalfa may be seeded into a native or cultivated species pasture. In other cases, the newly established species will produce forage at a different time than the existing species, as when tall fescue is overseeded with a warm-season legume or grass, or bermudagrass is overseeded with a cool-season species.

The object of introducing legumes into a grass pasture is to enhance the protein and mineral concentration of the herbage, to improve yields by increasing nitrogen availability, and to decrease weed growth by increasing sod vigor. As early as 1941, annual lespedezas were recommended for upgrading pastures in Missouri. In 1939, field trials in Kentucky indicated that clover stands could be established in old sods if suitable management practices were followed. In 1947, work in Pennsylvania showed that legume seedbeds could be prepared in old Kentucky bluegrass pastures, while work on hill pastures in West Virginia showed that shallow tilling gives better clover stands than does plowing. In Kentucky, in 1959, seeding trials showed that the establishment of alfalfa in tall fescue pastures is best achieved by a late winter or early spring seeding, that no-tillage establishments are as productive as plowed and disked areas, and that grass–alfalfa pastures yield more than tall fescue receiving 224 kg nitrogen/ha (200 lb/acre). Sods of species such as dallisgrass, bermudagrass, or bahiagrass can be cultivated with a one-way plow, a field cultivator, or a tandem disk. This will destroy about half the grass stand, but it gives satisfactory growth.

Thus, before 1960, it was shown that it was possible to renovate established grasslands. Since that time, two events have advanced renovation techniques. First, specialized tillage tools have been developed, and, second, herbicides are now available. Special sod-seeding machines are now able to prepare a narrow seed bed, lime, fertilize, and seed in one pass. Where herbicides are used, the object is to kill a limited portion of the grass sward. Usually about 25% of the grass is eliminated in bands 5 cm (1.8 in.) wide. This is satisfactory if seeding takes place while the grass is dormant. Many pasture drills have "spearpoint" openers which do not leave a rough seedbed. Planting equipment with "packing" wheels is also an advantage. Thus, in the United States, there is wide and varied experience of pasture renovation practices in the southeast and, to a more limited extent, in the north, west, and central areas of the country. For example, Table 6-7 shows the results of using two methods of seeding and several herbicide treatments in overseeding legumes into winter-dormant bahiagrass in Florida. Similar results have also been reported in Texas (Table 6-8), while in California there are indications of substantial increase in stocking rates from cow and calf herds (Table 6-9). In all cases, renovation has increased yields and considerably improved herbage quality. The chemicals listed in Tables 6-7, 6-8, and 6-9 (Paraquat, Dalapon Picloram, and Glyphosate) are those most commonly used in pasture renovation.

TABLE 6-7
Number of legumes 82 days after establishment in winter-dormant bahiagrass in Florida

	Seeding per m^2	Yield kg/ha	lb/acre
SEEDING METHOD			
Disk and broadcast	82	6,030	5,366
seeder	53	5,500	4,895
LEGUME SPECIES			
Alfalfa	102	8,160	7,262
Red clover	72	6,360	5,660
Ladino clover	33	2,760	2,456
HERBICIDE TREATMENT			
Paraquat before seeding and burnt	99	7,310	6,506
Dalapon	57	6,270	5,580
Paraquat at seeding	55	4,770	4,245
No herbicide	49	3,790	3,373

(Source: Kalmbacher et al., 1980, Agron. J. 72. By permission of The American Society of Agronomy, Inc.)

TABLE 6-8
Production rate and protein concentration in coastal bermudagrass following herbicide applications at 2.2 kg/ha (2 lb/acre)

Herbicide	kg/ha/day	Production rate lb/acre/day	Protein content %
Picloram	11	9.8	8.6
Tebuthiuron	37	33	13.0
Glyphosate	81	72	8.7
2,4-D	86	76	8.9

(Source: Baur et al., 1977, Agron. J. 69. By permission of The American Society of Agronomy, Inc.)

Work in Mississippi has developed a non-mechanical system for applying herbicides to weeds that are taller than the forage crop in which they are growing. The applicator requires no pumps or moving parts. A rope which acts as a wick to draw up the herbicide is mounted on a frame in such a way that it will rub across the taller weeds when moved over the plot. The selective control of johnsongrass in soyabeans using glyphosate has been successfully accomplished. It is also possible to drip feed Round-up onto a felt roller which contacts tall growing weeds as it is moved through the crop.

From studies of this type, certain generalizations have been developed. First, close grazing or frequent clipping will weaken the old sod, reduce its vigor, and help to provide suitable conditions for seedling development. Second, if the soil is at all acid, lime should be applied to bring the pH to a value suitable for legume growth. Third, the seeding equipment should be such that it will provide satisfactory, but not excessive, seed coverage. Fourth, the legume seed should be inoculated just before seeding. Fifth, the fields should be closely grazed while the legume plants are below the bite level of the animals. Given these considerations, pasture renovation appears to hold great promise for the improvement of forage yields.

TABLE 6-9
Summary of animal performance on improved and unimproved pasture in California, 1974–1975

	Unimproved	Improved
Animal unit months/ha	2.2 (5.4/acre)	3.3 (8.1/acre)
Cow		
Weight at calving, kg	381 (838 lb)	468 (1,030 lb)
Weight at weaning, kg	460 (1,012 lb)	478 (1,051 lb)
Calf		
Birth weight, kg	32 (70 lb)	33 (73 lb)
Weaning weight, kg	200 (440 lb)	197 (433 lb)

(Source: Raguse et al., 1980, Agron. J. 72)

FURTHER READING

Ayeke, C. A., and C. M. McKell, 1969, Early seedling growth of Italian ryegrass and smilo as affected by nutrition, J. Range Manage. **22:**29–32.

Blaser, R. E., W. L. Griffeth, and T. H. Taylor, 1956, Seedling competition in compounding forage seed mixtures, Agron. J. **48:**118–123.

Buxton, D. N., and W. F. Wedin, 1970, Establishment of perennial forages I. subsequent yields II. subsequent root development, Agron. J. **62:**95–97.

Chapin, J. S., and F. W. Smith, 1960, Germination at various levels of soil moisture as affected by ammonium nitrate of potash, Soil Sci. **89:**322–327.

Kilcher, M. R., and D. H. Henriches, 1960, Use of cereal grains as companion crops in dryland forage crop establishment, Can. J. Plant Sci. **40:**81–93.

Kittoch, D. L., and A. G. Law, 1968, Relationship of seedling vigor to respiration and tetrazolium chloride by germinating seeds, Agron. J. **60:**286–288.

Lawrence, T., 1963, A comparison of methods of evaluating Russian wildrye grass for seedling vigour, Can. J. Plant Sci. **43:**307–312.

Moline, W. J., and H. R. Robison, 1971, Effect of herbicides and seeding rates on the production of alfalfa, Agron. J. **63:**614–66.

Sund, J. M., G. P. Barington, and L. S. Scholl, 1966, Methods and depth of sowing forage grasses and legumes, Proc. 10th Internal. Grasses Congr. (Finland), Sec. 1:319–322.

Williams, R., 1964, The quantitative description of growth. *In* Grasses and Grasslands, Macmillan, New York, pp. 89–101.

Younger, V. B., and C. M. McKell, eds., 1972, The Biology and Utilization of Grasses, Academic Press, New York.

Chapter 7

The Physiology of Forage Crop Growth: The Grasses

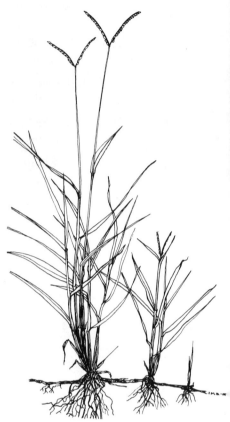

PLANT FOODS

Plants occupy a unique position in the world food chain, on which animals (including people) depend. The essence of their role lies in their ability to use simple inorganic substances to build complex organic compounds. The inorganic substances may be divided into two groups on the basis of their proportionate use. These groups consist of the major and the minor mineral nutrients (Table 7-1). To complete the list of inorganic substances required for plant growth, we must also add water, oxygen, and carbon dioxide. The organic compounds which the plant produces are used as a source of energy for the plant's vital processes, including growth and tissue repair. Such organic compounds may be regarded as plant foods essential for plant growth and development, as well as for the winter survival of perennial plants grown in the temperate regions of the world, and the withstanding of stress in other areas.

All living cells assimilate carbon dioxide, but differ as to the source of energy they use during the process. In the case of the green plants, radiant energy is absorbed and transformed into chemical energy as a part of photosynthetic activity. In chemical terms, this light process takes place in two steps. First, photo-oxidation of water splits off hydrogen and releases oxygen. At this stage, light is essential and temperature unimportant. This reaction supplies electrons for noncyclic phosphorylation. Second,

TABLE 7-1
Mineral nutrients required for plant growth

MAJOR OR MACRO NUTRIENTS	MINOR OR MICRO NUTRIENTS
Nitrogen	Iron
Phosphorus	Manganese
Potassium	Boron
Calcium	Copper
Magnesium	Zinc
Sulfur	Molybdenum
	Fluorine

in the dark reaction, which is temperature sensitive but does not require light, hydrogen is transferred to the carbon dioxide molecule and is bound as a metastable carbon compound. The chemical and energy balance of photosynthesis may be set out in the well-known formula

$$6CO_2 + 12H_2O \xrightarrow{2,828 \text{ kJ (675 kcal)}} C_6H_{12}O_6 + 6O_2 + 6H_2O$$

The carbon compounds produced in this way are utilized by all organisms as an energy source. This is true whether the organism is autotrophic (i.e., able to synthesize its own carbon compounds) or not. In the presence of elemental oxygen, the energy in the carbon compound is completely dissipated, hydrogen and oxygen recombine to give water, and the carbon dioxide which has served as a hydrogen carrier is released. The overall equation for this reaction summarizes respiration and may be regarded as the reversal of photosynthesis:

$$C_6H_{12}O_6 + 6O_2 \rightarrow 6CO_2 + 6H_2O + 2,828 \text{ kJ (675 kcal)}$$

This, and the previous simple formula, are the essential "statements" of life on this

TABLE 7-2
Carbohydrate substances most commonly found in green plants

Monosaccharides ($C_6H_{12}O_6$):
    Glucose and fructose

Disaccharides ($C_{12}H_{22}O_{11}$):
    Sucrose and maltose

Polysaccharides ($C_6H_{10}O_5)n$:
    Starch (glucose polymers) including amylose and pectin

    Fructosan (fructose polymers) including inulins and levans

146 / The Physiology of Forage Crop Growth: The Grasses

planet. "Life" has, in fact, been defined as the "heat" generated between photosynthesis and respiration!

The carbon compounds which are most commonly found in forage plants are set out in Table 7-2. Those substances with a simple molecular structure, such as the monosaccharides and disaccharides, are usually present in small amounts and may be regarded as metabolic intermediates. Either they are being used directly as energy for metabolism and the formation of plant tissue or they are in the process of being built up into the more complex polysaccharides. Where supplies of these simple sugars are inadequate, they may become limiting to growth and development in much the same way as water, temperature, or the major and minor inorganic plant nutrients.

CARBOHYDRATE RESERVES

Carbohydrate substances are needed not only for continued growth and development, but also for the production of polysaccharides. These polysaccharides form a reserve of carbohydrate called the *total available carbohydrates* (TAC). From this labile pool, the plant may draw material to offset both major and minor fluctuations in the production of simple sugars. As a result, limitations to growth and development, which a shortage of simple sugars might well produce, are avoided. The minor fluctuations in the production of simple sugars may be the outcome of periods of dull or cold weather, when photosynthesis does not proceed at the optimum rate or stops entirely. There are

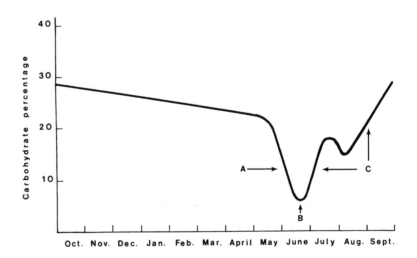

FIGURE 7-1
Annual fluctuation in carbohydrates reserves in smooth bromegrass in Alberta. (A) Spring growth reduces carbohydrate level; (B) New photosynthetic area supplies enough sugars for continued growth; (C) Carbohydrate reserves replenished.

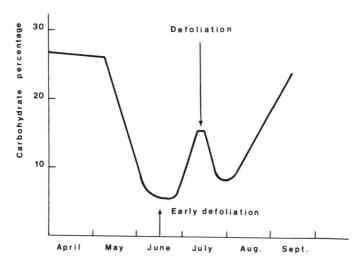

FIGURE 7-2
Annual fluctuation in carbohydrate reserves in smooth bromegrass.

also diurnal fluctuations in the production of simple sugars. During the hours of darkness, respiration continues and photosynthesis ceases, so sugars increase during the day and decrease at night. In the case of bromegrass, about one-third of the total available carbohydrate substances accumulated during the daylight hours is used during the night. Diurnal fluctuations are not so marked in most other species, but they are always present.

Major fluctuations in carbohydrate reserves are the outcome of respiration during the winter months, the production of new growth in the spring, and regrowth following defoliation. During the winter months, respiration continues at the expense of the plant's stored carbohydrate material. Hence, the total available carbohydrates (TAC), also known as the total nonstructural carbohydrates (TNC), are slowly reduced (Figure 7-1). Following the initiation of spring growth, the rate of depletion of the stored carbohydrates increases rapidly [Figure 7-1 (A)]. This marked withdrawal from reserves stops once the plant has developed a photosynthetic area which is capable of producing an adequate supply of sugars for shoot growth. [Figure 7-1 (B)]. Further increases in the photosynthetic area give a production of sugars sufficient to replenish the depleted reserves [Figure 7-1 (C)]. Carbohydrate reserves reach a peak in late July in Alberta, Canada. The decline in total available carbohydrates in August is the outcome of seed formation and leaf shedding. The subsequent buildup of carbohydrate reserves in late August and September is vital to the winter survival of the plant, since it must, at the beginning of the winter, have adequate reserves for both respiration during dormancy and regrowth in the spring. Without such reserves, winterkill is inevitable. In the warmer southern areas of the United States, such reserves are equally important for summer dormancy.

The effect of a single defoliation on carbohydrate reserves in smooth bromegrass is shown in Figure 7-2. It is important that reserves are permitted to increase prior to defoliation so that an adequate supply of carbohydrate material is available for subsequent regrowth. Had the defoliation illustrated in Figure 7-2 taken place earlier (mid-June), the carbohydrate reserves would not have been adequate for regrowth and the plants might well have died.

STORAGE REGIONS

Carbohydrate material may be stored temporarily in all parts of the plant. For longer-term storage, the stolons, corms, rhizomes, and stem bases are the principal storage organs in the grasses. While grass roots degenerate following defoliation, this would appear to be the outcome of a reduction in tiller leaf area and total available carbohydrates, rather than the utilization of root reserves for regrowth. This view is supported by evidence that, after a severe defoliation, only one-tenth of the carbohydrate material in orchardgrass roots is used for respiration or regrowth. It is possible that, after a severe defoliation, proteins from the roots are broken down and used for root respiration. However, it seems that in the grasses little or no root material is used to sustain the regrowth of the aboveground parts.

RELATIONSHIPS BETWEEN STORAGE SUBSTANCES AND SPECIES

The nature of the carbohydrate reserves varies from species to species and may be used as a basis for dividing plants into two types. Tropical and subtropical grasses accumulate starch. This substance is a pure glucose polymer (Table 7-2) which is present in two forms: amylose and amylopectin. The amylose molecule is linear and consists of between 50 to 1,500 glucose units. This substance is very soluble in hot water and forms 10% to 30% of most starches. The remaining 70% to 90% consist of amylopectin, a highly branched molecule, insoluble in water and with a high molecular weight (2,000 to 220,000 glucose units). An exception to the usual proportions is found in vegetative parts of corn (*Zea mays* L.), where amylose and amylopectin are present in about equal proportions.

Cultivated grasses grown in the temperate parts of the world accumulate fructosans. Native North American grasses are exceptions to this general rule in that they accumulate starches. This difference is taken to indicate that the North American grasses evolved in the southern parts of the continent, moving north when the Great Plains and prairie developed as a consequence of the upthrust of the mountain ranges on the west side of the continent (see Chapter 2).

The fructosans also occur in two forms: inulins and levans. Inulins are found mainly in dicotyledonous species, while levans from the major part of the fructosans found in the grasses. Levan molecules are linear and are highly soluble in cold water. In the tribe *Aveneae*, the linear molecules are present in long chains; those found in timothy, for example, consist of 260 fructose units. This contrasts with the tribe *Hordeae*, in which short-chain molecules are most common; Russian wildrye has a levan molecule that consists of 26 fructose units. The tribe *Festuceae* has species in which both long- and short-chain fructose molecules are present; the genus *Bromus* has short-chain fructosan molecules, while the genus *Dactylis* has long-chain molecules.

MANAGEMENT AND CARBOHYDRATE RESERVES

The level of the carbohydrate reserves in pasture grasses is an important management consideration. Temperature, soil moisture, and soil nutrients all influence carbohydrate reserve levels and consequently should be given careful consideration in management decisions.

Temperature and Carbohydrate Reserves

The optimum temperature for growth and net photosynthesis depends on the origin of the grass species. The main carboxylating enzyme in the temperate (C_3) grasses is ribulose-1, 5-diphosphate carboxylase, which has an optimum operating temperature of 20° to 25°C (68° to 77°F). The tropical grasses have both the C_3 (Calvin) and the C_4 (Hatch and Slack) photosynthetic pathways. The optimum temperature for growth and net photosynthesis for the C_4 pathway, where phosphoenol pyruvate carboxylase is the carboxylating enzyme, is 30° to 35°C (86° to 95°F). The site of the C_4 pathway is in the chloroplasts of the mesophyll tissue, while the site of the C_3 pathway is in the chloroplasts of the bundle sheath tissue.

The way in which temperature changes influence carbohydrate reserves is, then, dependent on the origin of the species concerned and the relationship between the environmental temperature and the photosynthetic optimum temperature. Thus, if the environmental temperature is below the optimum, a temperature increase will result in higher photosynthetic rates. This, in turn, might lead to increased carbohydrate reserves, depending on the use of photosynthates for growth. If the environmental temperature is optimal for photosynthesis, then change (either an increase or a decrease) could reduce carbohydrate reserves or stop their accumulation. An increase in temperature will always increase the respiration rate, possibly leading to a reduction in carbohydrate reserves. Hence, the effect of an increase in daytime temperatures may be very different from that of an increase in night temperature. High night temperatures frequently decrease reserves. Examples of this are found in many species. A marked reduction of carbohydrate reserves occurs in timothy, smooth bromegrass, orchardgrass, and Kentucky bluegrass when night temperatures increase. This applies not only to forages, but to all plants. It is for this reason that it is normal practice to reduce greenhouse temperatures at night.

Water and Carbohydrate Reserves

Scientific literature contains numerous reports indicating that drought may both increase and decrease carbohydrate reserves. The explanation for this apparent contradiction lies in the way in which the relative rates of both photosynthesis and respiration are affected by drought in the various forage species. The degree of water stress and the growth stage during which it occurs also have a variable effect on carbohydrate reserve levels. For example, orchardgrass, under conditions of water stress, shows some decrease in respiration rate and a *marked* decrease in photosynthetic rates, thus reducing carbohydrate reserves. For bromegrass, however, water stress has little effect on photosynthetic rates, while respiration rates are depressed to the same extent as for orchardgrass. Under these circumstances, carbohydrate reserves increase. These facts alone could well explain why bromegrass is generally regarded as being hardier than orchardgrass.

Nitrogen and Carbohydrate Reserves

As with soil moisture, the influence of nitrogen on carbohydrate reserves is varied and complex. However, some general conclusions may be drawn. If other factors are favorable, low or moderate amounts of nitrogen increase carbohydrate reserves. Such nitrogen applications increase the growth rate, so the leaf area, the number of chloroplasts, and the photosynthetic rate are all enhanced. Consequently, additional carbohydrates are available for storage. Heavy dressings of nitrogen will decrease carbohydrate reserves when environmental conditions are favorable for plant growth. The fertilizer stimulates the synthesis of amides and amino acids, and the carbon skeletons for protein synthesis are withdrawn from carbohydrate reserves.

The most disastrous combination of conditions is the application of high fertilizer dressing and frequent defoliation at a time when temperatures are high and soil moisture is low. All four of these factors tend to reduce carbohydrate reserves. High temperatures increase respiration, low soil moisture reduces photosynthesis, and frequent defoliation calls for a large withdrawal from carbohydrate reserves to initiate growth. Few species can survive such harsh treatment.

GROWTH AND REGROWTH

One of the essential features of a forage plant is that it should be capable of regrowth if successive defoliations are imposed. Both initial growth following dormancy and regrowth following mowing or grazing show the sigmoid growth pattern common to many biological organisms (Figure 7-3). A period of very slow initial growth is followed by a period of rapid increase, eventually reaching a plateau. The beginning of the plateau at time C (Figure 7-3) is, in fact, the time when growth becomes limited by the light falling on the pasture. At that time, the pasture canopy will intercept 95% of the incident light. The leaf area which will prevent all but 5% of the light from reaching the soil surface is called the *critical leaf area* and is usually defined by a *critical leaf area index* (Table 7-3). The leaf area index (LAI) is the number of units of area of leaf per unit area of soil surface and is different for the various forage species. The penetration of light into the sward is determined by the size, shape, and angle of the leaves. Thus, for the interception of 95% of the light, grasses, which have long, thin

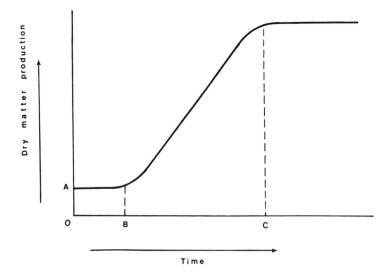

FIGURE 7-3
Generalized growth or regrowth pattern for forage plants.

leaves held at an acute angle, require a greater leaf area than do legumes, which have nearly horizontal, broad leaves (see Table 7-3).

As the grasses age and mature, the critical leaf area index will increase. The leaves of a young grass plant are held in a compact rosette. As the plant matures, the internodes elongate, the leaf arrangement becomes more open, and an increased leaf area is required to intercept 95% of the light falling on the plant. The reverse is true for legumes. Their broad leaves, which are nearly horizontal in the young plant, are entirely horizontal when the plant approaches maturity. The light reaching the soil surface decreases as the LAI is increased.

Another exceedingly important feature of the regrowth pattern of forage plants is the inverse relationship between their mass and the length of time which passes before growth becomes exponential. Figure 7-3 may be used to demonstrate this; there is an inverse relationship between the distance AO and OB. Severe defoliation of the grasses

TABLE 7-3
Critical leaf area index for forage species

	Critical LAI
Lolium perenne	7.1
Bromus inermis	6.9
Phleum pratense	6.5
Festuca rubra	6.3
Medicago sativa	4.6
Trifolium repens	3.5

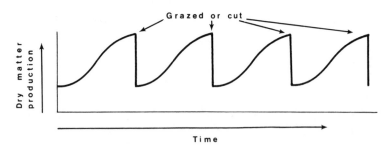

(A) LIGHT DEFOLIATION

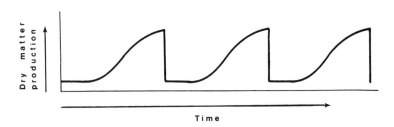

(B) SEVERE DEFOLIATION

FIGURE 7-4
Contrasting generalized patterns of regrowth for severe and light defoliation.

(i.e., where AO is small) will remove not only a substantial amount of dry matter, but also much of the carbohydrate reserve material stored in the stem base. Where carbohydrate reserves are low, initial growth will be slow (i.e., time OB will be long), and the total productivity of the stand will very likely be reduced. The objective of the pasture manager should be to put the plants into a series of sigmoidal growth and regrowth patterns (Figure 7-4). In doing so, the manager should endeavor to ensure that for the major part of the growing season the pasture has a growth rate of the type found between time B and time C in Figure 7-3. In other words, the height of defoliation should be such that OB is relatively short, as shown in Figure 7-4(A).

In Figure 7-4(B), severe defoliation has prolonged initial regrowth (OB in Figure 7-3), reduced the number of sigmoid regrowth cycles from four [Figure 7-4(A)] to three, and reduced the production from each cycle. These factors jointly contribute to an overall decline in pasture productivity.

A study of the University of Alberta data in Figure 7-5 shows not only that severe defoliation retards subsequent vegetative production, but also that the regrowth from the severely defoliated plants never equals that from a lightly defoliated plant; that is, the regression lines in the diagrams are parallel. This figure also indicates a similarity between the productivity levels achieved with heavy defoliation, when carbohydrate reserves are high, and those achieved with light defoliation, when carbohydrate levels are low. Furthermore, it is evident that the two species shown in Figure 7-5 respond differently to defoliation. No doubt the basal swellings found in orchardgrass store more carbohydrate material than the rhizomes of smooth bromegrass. Consequently, defoliations at the same height will slow initial regrowth of the bromegrass more markedly.

An equally important consideration in achieving maximum productivity from a given area over a given time is that the forage material should not be left unharvested after time C (Figure 7-3) has been reached. It is, however, a valid management decision to "store" (or stockpile) material in the field in this way, where circumstances so require (see Chapter 17).

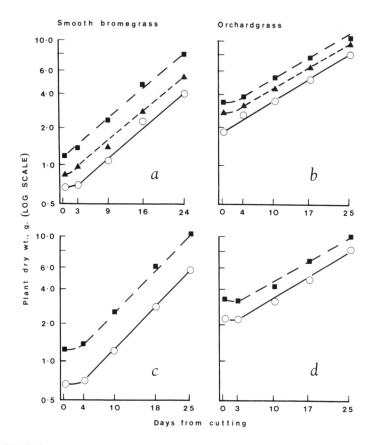

FIGURE 7-5
A comparison of regrowth rates for two grass species defoliated at three heights or with two carbohydrate levels. For (a) and (b), cutting heights were 2 cm (○), 5 cm (▲) and 10 cm (■). For (c) and (d), carbohydrate reserve levels were high, about 23% = (■), and low, about 10% = (○).

154 / The Physiology of Forage Crop Growth: The Grasses

DEFOLIATION OF A TILLER

The reasons for the inverse relationship between the defoliation height (OA) and the duration of the initial regrowth period (OB) are best explained by studying the effect of defoliation on a single tiller. Figure 7-6 shows a diagrammatic representation of an individual tiller which has fully developed and photosynthetically fully active leaves (*a*), leaves which are *partly* emerged from the sheaths of older leaves and have not yet reached their full photosynthetic potential (*b*), and leaves which have not yet emerged

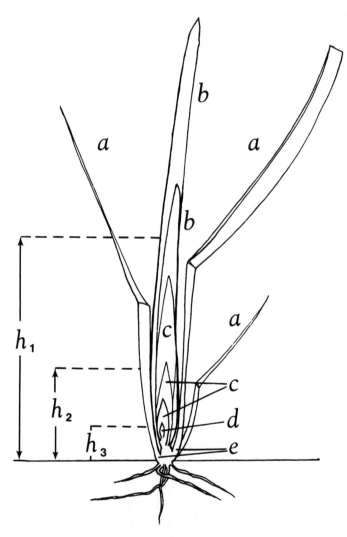

FIGURE 7-6
A nonelongated tiller. The key to the letters used is given in the text.

and are dependent on older leaves for the plant foods they require for further growth (c). Axillary buds (e) and the growing point or apical dome (d) are also shown. We will consider the effect of defoliating this tiller at three heights (h_1, h_2, and h_3). The actual dimensions of these three heights will vary with the size of the species. The important factor is not the actual height, but which organs are removed. At h_1, nearly all the most active photosynthetic areas are removed. At h_2, a substantial part of the leaf sheaths of almost all leaves are cut off. Cutting at h_3 removes the apical dome along with all the active photosynthetic areas.

The effect on the tiller defoliation at height h_1 will depend on growth conditions immediately prior to defoliation. Where conditions are favorable, and carbohydrate reserves are high, there will be little check in growth. Time OB will be short and rapid growth rates (BC) will soon resume (see Figure 7-3). Where nutrients (including reserves) are low, the delay in the resumption of growth will be much longer. Root growth will stop and there may be some deterioration of the root system.

Defoliation at height h_2 has a more drastic effect on the tiller. Almost all the photosynthetically active parts of the plant have been removed. The carbohydrate storage zone has also been depleted, since much of the stem base has been taken away. Under these circumstances, time OB (Figure 7-3) will be very much longer than when defoliation was at height h_1. Nitrates may accumulate in the lower part of the plant, since inadequate carbohydrate sources will prevent further protein synthesis. With inadequate carbohydrate material available for root respiration, root decomposition will take place on a substantial scale (Figure 7-7). Root respiration, root extension, and nutrient uptake from the soil are all substantially reduced. It is very important, when looking at a pasture, to keep in mind that there is as much plant material belowground as there is above and that defoliation will greatly influence both regions of the plant.

Finally, defoliation at height h_3 will remove the stem apex, which is the main growing point for the tiller. Under these circumstances, the tiller will die, but one or more of the buds in the leaf axil (e in Figure 7-6) will start to grow. Where floral development is taking place, the growing point may well have been carried up inside the column of leaf sheaths to a height greater than h_1. This is, in fact, the condition of most tillers when a crop is cut for hay.

Seasonal Changes in Growth Rates

Growth rates are not uniform within a single plant. Different parts of the plant achieve maximum growth rate at different times during the year. For the grasses found in temperate regions, there is a well-known inverse relationship between the vegetative growth above ground and the growth of roots, tiller buds, rhizomes, corms, and bulbs below ground or at the soil surface. During the cool weather in the early spring, when soil temperatures are below 10°C, root and tiller growth rates will be at a maximum. This below-ground activity decreases as the season advances, reaching a minimum at about the time when flowering takes place. With the onset of shorter days and cooler temperatures in the fall, the growth rates for underground organs will again increase. At this time, maximum growth rates will be achieved by rhizomes and stolons and by corms and bulbs. All these are organs in which the carbohydrate reserves for winter

FIGURE 7-7
Effect of defoliation on the root system of a grass plant: (A) not defoliated; (B) cut at height h_1; (C) cut at height h_2.

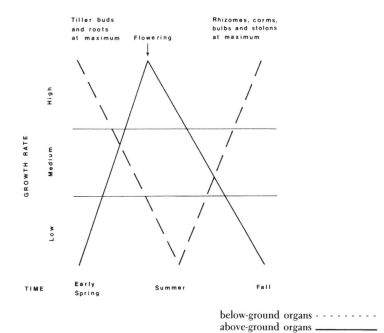

FIGURE 7-8
Seasonal changes in growth rates in grasses.

survival and spring growth will be stored. The above-ground growth rates reach a maximum in the early summer and are low in spring and fall (Figure 7-8).

WINTER DORMANCY

In the cool-season grasses, temperature and photoperiod are the primary factors which control winter dormancy. The photoperiod is of major importance in the induction of dormancy in such plants. The onset of short days (less than 13 hr) in the fall results in a sharp reduction in growth rates, and the plants enter a quiescent stage. This short-day-induced dormancy is closely associated with the plant's ability to survive the winter. There is evidence to indicate that phytochrome is the photoreceptor responsible for controlling dormancy. On the other hand, temperature is most important in breaking dormancy. As temperature increases in the spring, buds start to develop. Fluctuating day and night temperatures can also release dormancy.

There is evidence to indicate that temperature and photoperiod rarely act independently. At most stages of dormancy, each factor modifies the effect of the other. The plant response is complex and varies within species, having importance in the adaptation of a species over a range of environmental conditions. Other factors, such as mineral nutrient deficiency and lack of moisture, will induce the initial stage of dormancy.

SUMMER DORMANCY

In contrast with winter dormancy, summer dormancy, or heat resistance, is a response to high temperatures. In this case, the high temperature is frequently associated with a water deficit. There are marked differences between species, but in all cases prior exposure to high temperature greatly increases survival. Defoliation, in association with high temperature, will decrease survival. High temperatures and dry conditions increase carbohydrate reserves, while these reserves are decreased by defoliation.

There is some evidence to indicate that summer dormancy is determined in species like *Poa scabrella* by long days and high temperatures. However, there are cool-season grasses which enter the dormant state in response to high temperatures and low soil moisture only. Other grasses also in the genus *Poa* cease to grow even when kept well watered. For these latter species, summer dormancy is genetically determined.

PLANT STRESS

Legumes and grasses are exposed to many stresses. Cold, heat, wind, drought, shade, nutrient deficiency, ion toxicity, air pollution, and gaseous deficiencies are a normal part of a plant's environment. What, then, constitutes biological stress? According to Levitt (1972), "it is any environmental factor capable of inducing a potentially injurious strain in a living organism." These "potentially injurious strains" are opposed by the plant in two ways: (1) by the innate internal properties of the plant's physical and chemical response, and (2) by the plant's repair system. Furthermore, living organisms adapt or change gradually to decrease the strain caused by a stress.

Plants are exposed to two types of environmental stress: biotic and physicochemical. A fungal disease is an example of a biotic stress. Low temperature and air pollutants are examples of a physical and a chemical stress, respectively. The types of stress which a plant most commonly encounters are low temperature (cold stress), high temperature, water deficit, and high salt concentrations.

Low Temperatures

The biology of freezing temperatures is called *cryobiology*. It is a field of science that has developed in the last 20 years. For agricultural purposes, crops may be divided into those which are *tender*, or are killed by frost or low temperatures, and those which are *hardy*. The hardy species are able to avoid the formation of intracellular ice in their tissue or are tolerant of this ice formation. Hardy species prevent or reduce the development of intercellular ice by decreasing the total water content of their tissue, decreasing the free water in their tissue, and increasing their bound water content. *Bound water* is water which is held on the surface of protein colloids which form part of the protoplasm. The outcome of these changes is to decrease the viscosity of the protoplasm, to increase osmotically active solutes and the concentration of sugars, and to increase the permeability of the cell membranes. In short, hardy plants are able to withstand freezing because they are able to withstand freezing-induced dehydration.

High Temperatures

At first sight, it might seem that resistance to high temperature and drought injury would depend on the same mechanism as freezing tolerance, since dehydration plays a part in both processes. Indeed, there are similarities between freezing tolerance and thermotolerance. In both cases, there is a danger of cell membrane proteins unfolding. This process is likely to be irreversible and causes cell death if it occurs. In the case of the frozen plant, the membrane proteins are under tension; in the heated plant, the proteins unfold spontaneously if the temperature is high enough. The changes which are initiated in a plant in response to exposure to high or low temperature progress at different rates. The minimum time for low-temperature hardening is 24 hr, while high-temperature hardening is almost instantaneous. This is because high-temperature hardening involves a change only in protein conformation, while low-temperature hardening involves chemical reactions. The relationship between cold- and heat-hardening is such that cold-hardening has frequently been shown to increase drought resistance and heat tolerance, but drought and heat-hardening do not improve resistance to freezing.

Water Deficits (Drought)

In the vast, semiarid regions of the world, drought presents a serious problem for the agriculturalist. In such areas, native grasses, wheat, and sorghum are the most commonly grown species. Sorghum is a crop with an outstanding ability to initiate rapid regrowth following a period of drought. Research with this crop has shown that heat and drought stress are separate phenomena, since, for sorghum, drought affects photosynthesis by inducing stomatal closure, while high temperature has a direct effect on chloroplast activity.

While the total influence of drought on the plant's physiological process is not fully understood, for most plant species water stress will reduce translocation and nutrient uptake. Starch content of the leaves decreases and sugar content increases under drought conditions. In general, photosynthetic activities are reduced when the stomata are closed, but unexplained deviations from these general rules do occur. Low soil moisture does not affect photosynthesis and respiration in ladino clover. It is also of interest that there are many examples of growth being stimulated after release from stress caused by a moderate moisture deficiency. This could have important implications for irrigation techniques.

High Salt Concentrations

The forage species with resistance to saline soil conditions are dallisgrass, sudangrass, seaside bentgrass, and tall wheatgrass. The exact way in which saline conditions affect plant growth has yet to be determined. However, some aspects of the influence of salinity on germination, vegetative growth, and seed development have been studied. Plant reaction to salinity during germination is frequently very different from that during subsequent growth. For example, corn and many beans are salt-sensitive, but they are able to germinate at higher salt concentrations than those at which the salt-tolerant forage species listed above will grow. Such plants appear to have developed an "avoidance" mechanism. Salt accumulates in the upper layers of the soil where the seeds are placed, so normally plants germinate under salt concentrations higher than those in the rest of the soil profile. A fast-growing root system is able to grow away from, and so "avoid," saline conditions.

Salinity retards vegetative growth and reduces plant size and forage yield. In general, species or varieties which give high yields under normal conditions will give high yields under saline conditions. The exceptions to this general rule are of special interest. Studies of smooth cordgrass (*Spartina alterniflora*) have shown considerable success in isolating plants which grow vigorously in saline areas that are flooded daily. The forage production from this species is greater on poorly drained than on well-drained soils. These poorly drained areas have an abundant growth of blue-green algae.

It is important that further research into plant stress problems be undertaken. As world populations increase, the land areas devoted to the production of row crops will increase, and forage production will move to regions where cold, drought, heat, and salinity stress are commonly encountered.

FURTHER READING

Barnard, C., ed., 1964, Grasses and Grasslands, Macmillan, New York.

Dinauer, R. C., ed., 1969, Physiological Aspects of Crop Yield, Madison, Wisconsin.

Keller, W., and T. S. Ronningen, eds., 1975, Forage Plant Physiology and Soil-Range Relationships, American Society of Agronomy, Special Publication No. 5, Madison, Wisconsin.

Milthorpe, F. L., and J. D. Ivins, eds., 1966, The Growth of Cereals and Grasses, Butterworths, London.

Milthorpe, F. L., and J. Moorby, 1979, An Introduction to Crop Physiology, Cambridge University Press, Cambridge.

Zelitch, I., 1971, Photosynthesis, Photorespiration and Plant Productivity, Academic Press, New York.

chapter 8

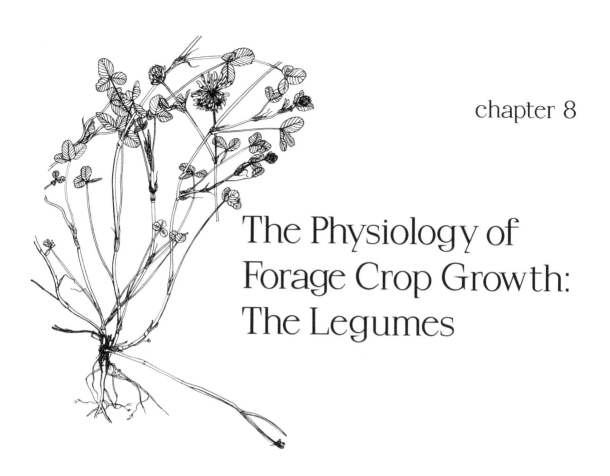

The Physiology of Forage Crop Growth: The Legumes

INTRODUCTION

All our crops consist initially of propagules (often seeds), which vary in size, nature, and number. The seeds which the farmer uses are genetically diverse and contain a food supply which is neither consistent in amount nor type. Such genetic variability is further modified and differentiated by the environment to provide even greater phenotypic diversity. All these varied individual plants which make up an established stand have in common their basic physiological processes. Each individual is a self-expanding unit which will use its production to increase its manufacturing potential (i.e., to increase respiration and photosynthetic rate), or its size and structure, or to add to its reserve of stored material. The total net production from the individual plants will depend on their number and the rate at which they work. Thus, the complexities involved in production from any growing crop, including forages, depend on a diverse group of organisms engaged in common processes. With this in mind, it is evident that much which was set out in the previous chapter will apply not only to the grasses but also to the legumes. Our objective here is to contrast the two plant types and to draw attention to minor variations in identical basic processes.

SHOOT GROWTH

The fundamental difference in morphology between the legume and the grass shoot is that the growing point of the legume plant is at the tip of an elongated stem which is frequently elevated. Consequently, with the exception of prostrate plants like white clover, when the legume plant is cut or grazed, the growing point is removed and growth from it is terminated. This is, in fact, the condition for all shoots in the spring, when low temperature or dry conditions have imposed a "dead" season. Using alfalfa as an example, let us now consider how an established legume plant would initiate new growth following such a period of dormancy.

The taproot of alfalfa is a thick, fleshy storage region for carbohydrate reserve material. The top of the taproot, called the *crown*, carries a large number of dormant buds. Five, six, or seven of these dormant buds will start to develop in the spring, using the carbohydrate reserve material stored in the root. The use of the root as a storage area contrasts with the grasses, in which reserve material is held in stem bases.

Just when growth is initiated in the dormant bud is not certain, but there is an increasing amount of evidence to indicate that it takes place much earlier than was at one time believed. Work at Beaverlodge in northern Alberta has shown that, in some years, buds which emerge aboveground in May actually start growth in December. As the soil warms, the buds emerge and grow, eventually producing a large enough photosynthetic leaf area to make them independent of the root reserve material. The regrowth pattern followed by the legume is the same as the sigmoid growth curve for grasses shown in Figure 7-3.

The number of stems produced in this way is closely associated with the overall productivity of the stand. The plant is able to adjust the number of stems it produces quite remarkably to compensate for high or low plant density. Air temperature, light intensity, soil moisture, and soil fertility influence plant vigor and, hence, stem number. These external factors influence the production of internal growth regulators which limit stem production. These regulators release apical dominance and increase the number of stems per plant.

The environment has a considerable effect on leaf growth. The shading of lower leaves in the canopy results in a higher leaf area per unit leaf dry weight than for fully exposed leaves at the top of the same canopy. Leaves produced under full incident radiation have more and larger palisade and mesophyll cells than those grown under shade. Temperature also influences leaf morphology. Leaves produced at the beginning and at the end of the growing season, when temperatures are *lower,* are *larger* than those produced under high temperature conditions in midsummer.

As the young shoot continues to grow, it becomes sensitive to photoperiod. Depending on the cultivar, day length greater than 13 hr causes the production of floral, rather than vegetative buds, in the leaf-stem axil. The onset of flowering initiates the release of apical dominance. Additional buds on the crown may start to develop, but more commonly it will be the vegetative buds on the lower parts of the shoot that will produce subsequent vegetative growth. The height of cut will consequently determine the number of buds available for regrowth. The height of the remaining stubble will

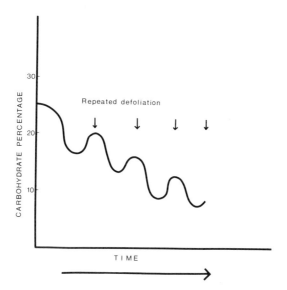

FIGURE 8-1
Decline in carbohydrate reserves associated with the repeated defoliation of alfalfa shoots. The plant is defoliated again before it is able to replace the reserves used in producing the initial growth following an earlier defoliation.

also determine the proportion of the plant's labile pool of carbohydrates left behind after cutting. On both accounts, a high cut will enhance regrowth.

If the plant is cut just after flowering starts, the buds which form the regrowth are in a suitable physiological condition to initiate new vegetative growth. It is for this reason that it is commonly recommended that alfalfa be cut for hay at the early bloom stage. If the plants are cut earlier, there is a delay in the initiation of regrowth, while the auxin mechanism is reversed. Then the overall production in time (i.e., the growth rate) is reduced.

It is important that the young shoot *not* be severely defoliated before it has had the opportunity to replace and build up the carbohydrate reserves withdrawn for its initial growth. This is likely to occur under conditions where continuous grazing is practiced. The young alfalfa plant is attractive to cattle, especially if the grass in the pasture has passed the juvenile stage. Under these circumstances, the animals will return repeatedly to graze the alfalfa plant as soon as it produces new regrowth. The carbohydrate reserves are then depleted with each defoliation, and the plant becomes incapable of regrowth and dies (Figure 8-1). This situation is very similar to that shown for the grasses in Figure 7-2, but the number of growing points below grazing height for the legumes is much fewer than for grasses.

ROOT SYSTEM

The root system, which we have already discussed as a storage organ, will now be studied in terms of its ability to maintain vigorous herbage production. The role of the legume root system in absorption and translocation is essentially the same as that of any other dicotyledenous plant. We should, however, consider the effect of defoliation on the legume roots.

In general, legumes have deep root systems and are capable of obtaining moisture and nutrients from much greater depths than the grasses. When the legume plant is defoliated, the root system will not degenerate in the manner shown for the grasses in Figure 7-7. There is little loss in root weight, but the loss in absorbing area is substantial. The extension of root tips ceases, and root hairs die and are not replaced. Under these circumstances, a root with a diameter of 0.33 mm, which carries root hairs 2 mm in length, will have its absorbing area reduced 150-fold when all its root hairs are shed. Until photosynthates from new leaves are available for the development of new root hairs, the intake of moisture and mineral nutrients will be substantially reduced.

The other serious consequence of defoliation for the legumes is that photosynthates are no longer available to provide the *rhizobia* in the root nodules with the sugar they require. The nodules cease to function and are shed into the soil. In a pasture composed of different species, nitrates liberated from the nodules may then become available to the grass. This is especially serious if the legume has been defoliated as a result of preferential grazing. The nitrates will then be readily taken up by the non-defoliated grasses, which are then able to produce additional growth and suppress the legume.

GROWTH RESPONSES

To disentangle the influence of complex physiological processes, the plant physiologist tries to isolate them into their component parts and study them individually. For the agriculturalist who is interested in the study of the whole plant, this division into distinct functions has certain dangers. There can, in fact, be no complete separation of these plant processes. Respiration influences photosynthesis and translocation and also determines the rate at which photosynthetic products and mineral nutrients may be utilized. Rates of both respiration and photosynthesis vary with leaf water potential and are influenced by the availability of mineral nutrients and plant food. While plant physiologists have elucidated many of these component parts in considerable detail, and while there is a reasonable understanding of the way they fit together, certain important features remain unknown. We cannot yet assess the interactions of component physiological processes with sufficient accuracy to predict the behavior of the whole plant with certainty. Until advances in our understanding permit such predictions, we must continue to examine, with some caution, the way in which the individual plant processes are influenced by environment and the effect that this has on plant growth.

Influence of Light. Both the quality and the quantity of light influence photosynthetic responses in the legumes. The leaves use radiant energy in the range from 400 to 700 nm; about 80% of the radiation within these wavelengths is absorbed. Light intensity required for light saturation varies with species and winter temperature. For example, red clover and birdsfoot trefoil reach light saturation at relatively low light intensities when temperatures are high (30°C or 85°F), while alfalfa, grown at the same temperature, continues to respond to increased light intensity at a level more than double those at which responses cease for the other two legumes. For many legumes growing under field conditions, a reduction of full sunlight (7,000 to 12,000 ft-c) by 50% to 75% has little effect on either stem or root growth. There is substantial evidence to show that light intensities below 3,000 to 4,000 ft-c will limit carbon dioxide uptake, so that light intensities below saturation levels will occur on cloudy days, in the early morning, and in the evening. High specific leaf weight in alfalfa has been shown to be associated with high photosynthetic capacity. Considerable variation exists among alfalfa cultivars for both these characters, so genetic selection for either or both traits might lead to yield improvement.

In general, alfalfa and most other legumes have a photosynthetic capacity below that of the tropical grasses. Ribulose-1, 5-diphosphate carboxylase is the main carboxylating enzyme in legumes, rather than phosphoenolpyruvate carboxylase, which is present in the tropical grasses and a few dicotyledonous plants.

Canopy Structure. Light penetration to the soil surface shows a logarithmic decrease as the leaf area index (LAI) increases. The leaf area index is the number of units of area of leaf per unit area of soil surface (see Table 7-3). In contrast with the grasses, the critical LAI for legumes decreases as the plant ages, the leaves of the legumes being held horizontally in the autumn (0° to 30° from the horizontal) and somewhat less so by the young growth in the spring (30° to 60°).

For the legume, diurnal changes also exist in orientation of leaves in the upper part of the canopy. This orientation is with respect to the azimuth. The leaflets will move through as much as 140° per day as the leaf surface is held toward the sun. During the summer in the northern hemisphere the majority of the leaves will face north. Lower leaves in the canopy show little movement of this type.

To study crop canopy structure, a number of other crop parameters are calculated. Multiplying the soil area occupied by the crop by its height will give the canopy volume. This value is then divided by the LAI to give the *leaf area density* (often called *Lv*). The leaf area density varies for different crops and for different heights within the same crop and is also influenced by environment. Of the total canopy volume, 1% or less is occupied by plant tissue.

The nature of the canopy architecture influences both photosynthesis and evaporation. It is the visible radiation which is of importance for photosynthesis, while the net radiant energy (total incoming short-wave radiation less the outgoing long-wave radiation) is available for evaporation.

Leafage. The photosynthetic activity of the legume leaf declines with age. There is an increase in CO_2 fixation during the first 5 to 10 days after a leaflet is fully expanded, after which photosynthesis declines until senescence. The rate of this decline is much influenced by environmental conditions. Termination of photosynthesis usually takes between 30 and 60 days, but may be much shorter (21 days) in species like white clover.

There is no doubt that the shading of lower by upper leaves plays an important part in causing a rapid decline in photosynthesis. There is, however, little indication that older, shaded leaves parasitize the plant, since respiration rates are much lower than those found in upper leaves, and the bottom leaves of the canopy do not receive photosynthates from those at the top.

Respiration. The plant's energy balance is greatly influenced by the water loss which occurs during respiration. For example, high temperatures can depress net CO_2 uptake by one-third. Root respiration plays an important part, which is frequently overlooked, in influencing the plant's energy balance. Respiration from plant tops and roots is about equal under normal growing conditions. Root respiration is markedly and rapidly reduced after cutting. It increases again gradually as the plant regrows. This suggests that legume roots first use synthesized carbohydrates, and not the carbohydrate material which they store. Root reserves appear to be preferentially used for the production of new top growth.

Soil Moisture. Both the respiration and photosynthetic processes in the legumes are much less influenced by water shortages than in many other species. For example, in alfalfa, photosynthesis starts to decrease when soil moisture is reduced to 35% of the soil's maximum water-holding capacity. This contrasts with most other crops, where photosynthesis decreases at 45% to 55% of the soil's maximum water-holding capacity. The respiration of red clover is influenced in a similar manner and does not change appreciably until soil moisture declines to 25% of maximum. In the case of alfalfa, this apparent drought tolerance is independent of root depth, a valuable asset under some field conditions. Some legumes, including alfalfa, are sensitive to flooding. Eight days of flooding has been shown to reduce photosynthesis in alfalfa by 30% under circumstances where orchardgrass, ladino clover, and crimson clover were not affected. (See Table 4-1.)

FURTHER READING

Bolton, J. L., 1962, Alfalfa, Leonard Hill Ltd., New York.

Bray, J. R., 1963, Can. J. Bot. 41:65–72.

Bula, R. J., and D. Smith, 1954, Agron. J. 46:397–401.

Calder, F. W., and L. B. MacLeod, 1966, Can. J. Plant Sci. 46:17–26.

Cooper, C. S., and C. A. Wilson, 1968, Crop Sci. 8:83–85.

Hanson, C. H., 1972, Alfalfa Science and Technology, American Society of Agronomy, Madison, Wisconsin.

Hatch, M. D., and C. R. Slack, 1966, Biochem J. 101:103–111.

Hunt, L. A., C. E. Moore, and J. E. Winch, 1970 Can. J. Plant Sci. **50**:469–474.

Pearce, R. B., R. H. Brown, and R. E. Blaser, 1968, Crop Sci. **9**:423–426.

Robison, G. D., and M. A. Massengag, 1968, Crop Sci. **8**:147–151.

Smith, Dale, 1969, Agron J. **61**:470–473.

Thomas, M. D., and G. R. Hill, 1949, Photosynthesis in Plants, Iowa State University Press, Ames, Iowa.

chapter 9

Forage Quality

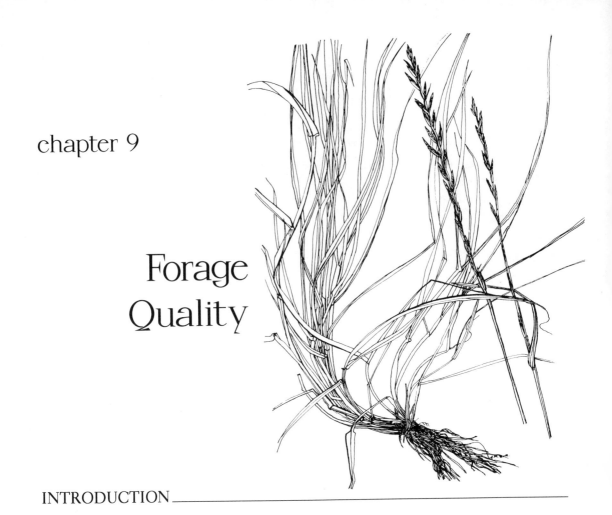

INTRODUCTION

Feed quality has been defined as the amount of nutrient material that an animal can obtain from a feed in the shortest possible time. This is an unduly simplistic definition, for the determination and study of quality in any crop presents many complex problems. Forages are no exception. In fact, the evaluation of quality in herbage crops provides many unique and interesting areas of study. These arise from the diverse ways in which forage quality may be measured. First, forage quality can be determined by chemical analysis, which aims to study those compounds (carbohydrates, proteins, minerals, and vitamins) on which animal nutrition depends. Second, it might be considered specifically in relation to the fiber content of the material. Many, but not all, of the different types of fiber may be digested at varying rates and utilized by the ruminant animal. Third, the extent and rate of digestibility of forage material is of the greatest importance in animal nutrition and is frequently studied to the exclusion of all other characters. Fourth, animal intake, which is closely associated with liveweight gain, is dependent on both the physical and the chemical nature of animal foodstuffs. Finally, the value of a feeding stuff may be determined entirely from its energy content, an important animal requirement.

TABLE 9-1
The principal constituents of grassland herbage

1. Water

2. Nitrogenous compounds
   a. Proteins
   b. Nonprotein compounds: peptides, amino acids, amides, purines, pyrimidines, nitrate, and cyanogenetic glycosides

3. Carbohydrates and pectin
   a. Nonstructural: hexoses, oligosaccharides, fructosan
   b. Structural: cellulose, hemicellulose, pectin

4. Lignin

5. Lipids
   Fats, waxes, phosphatides, sterols

6. Organic acids

7. Pigments: chlorophylls, carotenoids

8. Vitamins

9. Minerals

(Source: Worden et al., 1963, Animal Health, Production and Pasture)

FORAGE CHEMISTRY

Let us first study forage quality from a chemical point of view. We will examine the chemical substances which are found in forage plants and see how they may relate to the animal's nutrient requirements. The growing plant contains large amounts of water. The actual amount may be quite variable, ranging from 70% to 90%. For this reason, it is usual to express forage constituents on a dry-matter basis. Percentage protein, for example, is usually expressed in this way.

The nine principal groups of chemical constituents found in forage plants are given in Table 9-1. We will now consider these in the order listed in the table.

NITROGENOUS COMPOUNDS

Proteins are found in a soluble form in the plant's cytoplasm and in a granulate or particulate form in the chloroplasts. The proportions of these two types of protein vary from species to species. Most forage plants have 40% of their protein in the chloroplasts. This fraction is relatively indigestible. The remainder is located in the cytoplasm and is readily digestible. Because the protein molecules are very large, their chemical analysis presents a problem. Even when the cell has been killed, the cell wall will act

NONPROTEIN NITROGENOUS COMPOUNDS

Amino acids are the basic compounds from which proteins are formed. More than 20 compounds of this type are known, each occurring in differing proportions in different proteins. Ten of these are essential to the diets of farm animals, and two others are required by many animals (Table 9-2). For satisfactory nutrition, it is important not only that all the essential amino acids be present in the leaf herbage, but also that they be present in the appropriate proportions. If a given quantity of leaf protein contains only half the amount of an amino acid required for, say, milk production, then twice as much of the forage material must be ingested if that amino acid is not to become a limiting factor. There is evidence to indicate that lysine and tryptophan may both limit milk production in this way. Other nonprotein nitrogen compounds present in herbage are those concerned with protein metabolism: nitrates, ammonia, acid-amides, peptides, simple purine bases (such as stachydrine), choline, pyrimidines, and, in small amounts, cyanogenetic glycosides.

TABLE 9-2
Amino acids essential for animal nutrition

Arginine	Methionine
Histidine	Phenylalanine
Leucine	Threonine
Isoleucine	Tryptophan
Lysine	Valine

May also be required:
Cystine
Tyrosine

CARBOHYDRATES, INCLUDING PECTIN

Carbohydrate substances in the plant are divided into two major groups. Nonstructural carbohydrates consist of digestible sugars, which include monosaccharides, disaccharides, and polysaccharides (see Table 7-2). Structural carbohydrates (including pectin) are the substances which form the cell walls and those tissues which give rigidity to the plant. These include cellulose, hemicellulose, and lignin. The nonstructural carbohydrates were discussed in Chapters 7 and 8, where carbohydrate reserve materials were considered.

The main structural carbohydrate component found in plants is the cell wall substance cellulose, which may be digested by bacterial action in the rumen. This substance is a glucose-polysaccharide made up of long chains of glucopyranose units linked by β-glucosides through carbon atoms 1 and 4. The macromolecules so produced have a zigzag arrangement. Such chains are closely packed into aggregates called microfibrils. As the plant becomes older, the microfibrils are coated with lignin and hemicellulose. In general, the clovers contain a lower percentage of cellulose, on a dry-matter basis, than do the grasses.

Hemicellulose is a complex mixture of substances, including short-chain glucans, polymers of xylose, arabinose, mannose, and galactose, mixed polymers of sugar units and uric acid units, as well as residual pectin polysaccharides. Pectic substances are usually more plentiful in the legumes (about 8% of the dry weight) than in the grasses (about 1%).

Lignin

Lignin is a high-molecular-weight aromatic complex which is deposited, as maturation takes place, on the microfibrils and other cellulose structures in the secondary cell walls of plants. It will vary from about 2% of the dry matter in the young plant to 17% or more in a fully mature plant. Lignin contains methoxyl groups which vary in some species with the age of the plant. In pasture grasses, however, they remain reasonably constant. Lignin, and any cellulose structure which it coats, is not digestible by ruminant animals.

Lipids

The lipid substances found in forage plants consist of a complex mixture of fats, waxes, sterols, and phospholipids. The fatty acids of the true fats vary from 2% to 4% of the herbage dry matter. Those found in the leaves differ from both seed and animal fats in that they contain no oleic acid. The major components are linoleic and linolenic acids, which are present in about equal proportions. The maturation of forage plants is accompanied by a decrease in fatty acid content and a reduction in the proportion of unsaturated fatty acids. Hay making will also result in a large, if not complete, loss of unsaturated fatty acids.

Organic Acids

Organic acids are products from the earlier stages of photosynthesis and play an important role in the metabolism of fats, carbohydrates, and proteins. To perform their numerous functions, the organic acids are many and varied. Malic acid is the one most commonly found in herbage and, together with citric acid, makes up about 50% of the total organic acid present. Legumes contain more organic acid substances than do grasses.

Pigments

Two types of pigment are found in pasture herbage: chlorophylls and carotenoids. The green chlorophylls, which have no direct nutritional value for the grazing animal, do contain magnesium, which is an important element. These pigments are complex esters of tricarboxylic acid. However, because a bright green leaf color, indicating a high chlorophyll content, is usually associated with high-protein and high-carotene contents, it is used as a character in judging both of these forage qualities.

Vitamins

The carotenoids found in pasture plants are almost entirely β-carotene, the precursor of vitamin A. Since this is the only source of vitamin A available to the grazing animal, it is of considerable nutritional importance. Legumes have higher carotene content than do the grasses. For both species, the carotene content of the young plant is high, but decreases as the plant matures. Carotene is highly unsaturated and is consequently a very unstable compound whose oxidative destruction commences as soon as the herbage is cut. Hence, the carotene content of field-dried hay is usually negligible. However, high carotene content in stored forage is found only in artificially dried material such as alfalfa (see *dehydration* in Chapter 11).

By microbial activity in the rumen, ruminant animals can synthesize many of the vitamins they require. These include vitamin C (absorbic acid) and components of the vitamin B complex (thiamin, B_1, and riboflavin, B_2). These substances are also present in all green forage. Although vitamin D is not present in fresh green herbage, small quantities of ergosterol are present. This can be converted to vitamin D by ultraviolet radiation during hay making. Vitamin A, which is required for animal nutrition, is absent in fresh herbage, other than in the form of its precursor β-carotene. Vitamin E is present in grassland herbage.

Minerals

Because literature on the mineral content of forage plants is very extensive, it cannot be reviewed here in its entirety. In general, plant tissue reflects soil deficiencies. Some of these deficiencies may be corrected in the forage by fertilizer applications. Animals rarely suffer from a complete lack of the minerals they require, but partial deficiencies resulting from forage material with a low mineral content are common. Most legumes are rich in calcium, copper, potassium, and magnesium. Alfalfa may be deficient in phosphorus, a mineral important to livestock for skeletal development. Many herbaceous weeds are also rich in important mineral substances (Table 9-3), and they, too, will contribute to livestock health.

The plant and animal requirements for many minerals differ considerably. For example, phosphorus is limiting to plant growth at a much lower level than that required by animals. Thus, major animal disorders, and even death, may occur on pastures whose soil has amounts of phosphorus satisfactory for vegetative growth. Problems of this type are frequently overcome cheaply and easily by using salt and

TABLE 9-3
Principal mineral constituents of grasses, legumes, and herbaceous weeds (percentage)

	Calcium	Phosphorus	Potassium	Magnesium	Sodium	Chlorine
Grasses	0.39	0.21	1.99	0.24	0.13	0.50
Legumes	1.69	0.38	2.41	0.69	0.07	0.41
Herbaceous weeds	1.41	0.35	3.07	0.75	0.15	0.51

(Source: Worden et al., 1963, Animal Health, Production and Pasture)

mineral licks to meet the livestock requirements, especially when the deficient mineral is iodine, cobalt, or calcium.

In summary, it is evident that our forage material contains a considerable number of very complex compounds, some of which are important and others of which are not. How are we to analyze our forage material in such a way that we evaluate quality by means of one, or a few, parameters which will be comparable from one type of material to another? We will consider various ways in which this may be done in the remainder of this chapter.

PROXIMATE FEEDSTUFF ANALYSIS

The proximate feedstuff analysis is the oldest and most commonly used form of forage evaluation. It was developed in Germany in 1860 by Henneberg and, with modifications, has been used since then in many parts of the world. First, the percentage of moisture in the material presented for test is determined. The percentage of moisture commonly ranges from about 70% for a silage sample, down to 13% for a hay sample, and is as low as 5% for an artificially dried forage sample. All subsequent determinations are expressed on a dry-matter basis.

Crude Protein (CP)

Next, the percentage of *crude protein* in the forage material is determined. This is done by analyzing the herbage to find the proportion of nitrogen in the dried sample and multiplying the result by 6.25. The value so obtained is the *crude protein content* of the sample. The nitrogen determination may be made by using the Kjeldahl method or by using one of the many types of automated analysis equipment which are now available on the market. The value 6.25 is a generalized proportion of elemental nitrogen to plant protein, taking into account the fact that the forage material contains ammonia ions, nitrates, and amides. These nonprotein nitrogen substances can also be used by the rumen flora and fauna to build proteins. Ruminant animals require 8%

TABLE 9-4
Proximate feedstuff analysis of forage samples from Alberta, Canada

	No. of samples	Moisture %	Crude protein %	Fiber (ADF) %
HAY				
Alfalfa	117	13.8	17.4	38.1
Red clover	15	10.5	14.4	43.6
Bromegrass–alfalfa	324	12.0	13.5	38.9
Bromegrass	30	10.8	9.5	40.1
Timothy	16	10.2	8.1	37.1
Native grass	11	11.7	6.3	45.2
SILAGE				
Oats	28	67.5	8.4	38.0
Barley	36	64.1	9.7	33.8
Alfalfa–grass	31	62.8	14.7	40.5

(Source: Collected from various publications of the Alberta Soil and Feed Testing Laboratory and from unpublished data from the University of Alberta forage quality laboratory)

to 10% crude protein for maintenance and up to 15% crude protein in the case of a high-producing dairy cow.

The values obtained for crude protein give no information about the amino acids present and consequently cannot be used to evaluate protein quality. The digestibility of the crude protein in forages varies widely, ranging from 35% to 80%. The amount of nonprotein nitrogen in silage is usually higher than in other types of conserved forage, since some protein nitrogen may be converted to nonprotein nitrogen during the silage-making process (see Chapter 12). In view of its high digestibility, the percentage of protein in forage is frequently regarded as an index of digestibility. In general, leguminous plants have a higher protein content than grasses (Table 9-4).

Crude Fiber (CF)

One of the most important parameters, determined by the proximate feedstuff analysis, is the percentage of crude fiber in the sample. In the original method, crude fiber was regarded as the residue left after boiling successively with dilute sulfuric acid and sodium hydroxide. The intention was to separate the carbohydrate fraction of a feedstuff into crude fiber, the residue (which was assumed to be indigestible), and nitrogen-free extractive (NFE) components, which are removed by the treatment and determined by subtraction. In fact, many components of the NFE were less digestible than the cellulose which formed part of the crude fiber. For example, lignin, which is entirely undigested, dissolves in sodium hydroxide, and hemicellulose, some components of which are indigestible, dissolves in both acid and alkali. Consequently, all or part of these substances become components of the nitrogen-free extracts.

Acid-Detergent Fiber

In recent years, a number of workers, among whom P. J. Van Soest is the most prominent, have developed alternative methods. The most widely used of these is the acid-detergent fiber (ADF) determination. The forage material is boiled for 1 hr in a normal solution of sulfuric acid to which a detergent (cetyltrimethylammonium bromide) has been added. This treatment yields the percentage of lignin, cellulose, and silicon oxide in the sample so tested. An alternative method, neutral-detergent fiber (NDF), yields the percentage of total cell wall substances. Again the NFE (this time containing different components) may be obtained by subtraction.

The total carbohydrates (NFE + fiber) are estimated by subtracting the total of the percentages of ash, ether extracts, and crude protein from 100. The NFE, as previously explained, is the result of subtracting from the total carbohydrate fraction either the crude fiber percentage or the percentage of ADF. Ether is used to extract fats, waxes, chlorophyll, and volatile oils (all of which are called, collectively, the *ether extractives*). The ash is used to determine the percentages of phosphorus, calcium, and other minor elements in the feedstuff.

The criticism of proximate feedstuff analysis, even with its modern modifications, is that no component analyzed represents a single chemical substance (with the exception of water). Furthermore, the data are of limited value in determining digestibility, since the extent to which cellulose, hemicellulose, and total cell wall contents are digestible differs among plant species and varies with the plant's stage of maturity and the animal concerned.

_____ REVIEW

What, then, are the general quality characteristics of a forage? To this point, we have shown that forages contain a high proportion of fiber (over 25%), consisting of cellulose, hemicellulose, and lignin, and a low percentage of sugars, starch, and fats. They are frequently evaluated by using the proximate feedstuff analysis in which

$$NFE = 100 - \text{ether extractive} - CF \text{ (or ADF)} - CP$$

is the general relationship between the components studied. Forages, in fact, provide an important source of energy for the ruminant animal, a factor which we should now consider.

_____ ENERGY

Forage may be evaluated in terms of energy. Such considerations are important, because when forages alone are fed, energy may limit animal performance. The total daily energy intake may be determined in joules or kilojoules (kJ) (megacalories, Mcal) by burning the material eaten by an animal in a single day in a bomb calorimeter. For example, a steer eating alfalfa–bromegrass hay and making a small body-weight gain each day has a total daily energy intake that is divided into that which is passed out in the feces (47%) and the remainder, called the *digestible energy* (53%). Of the digestible energy, 5% is lost as urine and 6% in a gaseous form (mainly methane). The remaining 42%, the *metabolizable energy*, may in turn be divided into two parts: that which is used in achieving a liveweight gain (about 2%) and that which is used in heat production to maintain body function (40%). Higher energy feeds such as grain, which have a higher caloric density, have a greater proportion of the total energy available for weight gain relative to maintenance.

DIGESTIBILITY

Digestibility is frequently considered to be the most valuable estimate of forage quality, since it is closely associated with animal productivity. Digestibility may be related to dry matter, to energy, or to any component of the nutrient material available in the feed. Dry-matter digestibility is determined by collecting rumen fluid from an animal with a rumen *fistula* and conducting a standardized *in vitro* determination of dry-matter disappearance. This method is open to two types of experimental error. First, bacterial residues which would be digestible in the animal's lower tract accumulate and result in low digestibility values. Second, the rumen fluid is variable. It may vary with changes in the diet of the animal from which it is collected, with changes in the time of feeding and the time of sampling, with disease, and with the disposition of the animal used. Thus, the success of the method depends on careful standardization at all stages.

INTAKE

The ruminant animal has four stomachs: the rumen, the reticulum, the omasum, and the abomasum. For our purposes, the rumen is the most important. It may be regarded as a vast fermentation vat into which large amounts of saliva (80 liters or 16 Imperial gallons for a cow and 10 liters or 5 Imperial gallons for a sheep) are poured each day. The food is swallowed, mixed with saliva, regurgitated, chewed, and swallowed again. In the rumen there are a large number of microorganisms, from very many different species, which are able to digest cellulose and to synthesize amino acids and some vitamins.

The forage or roughage part of an animal's diet will distend the rumen and provide stimulation, which leads to the secretion of the large amounts of saliva required for the bacteria to function. A poor-quality forage (e.g., straw) will stimulate the rumen, but will not provide enough nutrients, whereas a concentrated feed (e.g., grain fed alone) will provide adequate nutrients, but will not fill and stimulate the rumen.

Intake is high when the material on which the animal is feeding is palatable (see Chapter 16). Figure 9-1 shows that the intake of a palatable forage (A) may be ten times that of poorer material (B). Palatability is not the only factor which determines intake.

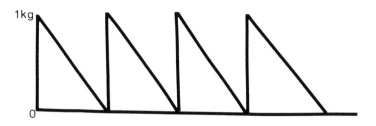

(A) GOOD-QUALITY FORAGE

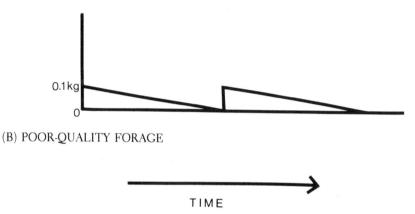

(B) POOR-QUALITY FORAGE

TIME

FIGURE 9-1
Intake and digestibility for good- and poor-quality forage.

Digestibility plays an important part. If material is readily and quickly digested, the animal will start to feed again and the total daily intake will be substantially increased (Figure 9-1).

The daily intake of forage material is about 2.5% of the animal's liveweight for a top-quality forage, 2% where the forage quality is good, and only 1.5% where the forage quality is poor. Keeping in mind these facts and the relationship set out in Figure 9-1, it is not surprising that the feeding value of a hay may be increased quite remarkably by chopping or grinding. The animal is able to consume more feed if the size of the pieces of which it is composed is small, since more material may be packed into the rumen.

Experiments conducted at Melfort, Saskatchewan, compared good-quality bromegrass hay (with 17% crude protein) with poor-quality stipa hay (7% crude protein). Both hays were fed in the long form (baled), as well as chopped, ground, and ground and pelleted. The relationship between intake and average daily gain is shown in Figure 9-2. This figure indicates that fewer pounds of smooth bromegrass hay were required for a pound of animal-weight gain than of the green stipa hay. Also, processing had a much greater effect on the feeding value of the poorer-quality forage.

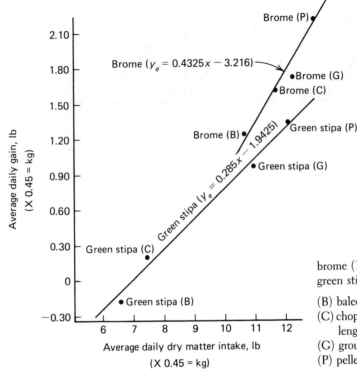

FIGURE 9-2
Effect of chopping, grinding, and pelleting on feed intake and liveweight gains of steer calves fed good and poor hay. (Source: Agriculture Canada Research Station, Melfort, Saskatchewan)

VARIATION IN FORAGE QUALITY

Forage, in terms of quality, is the most heterogeneous of our agricultural products. This fact is commonly overlooked, but can be illustrated by comparing forage and grain crops. First, there are many more forage crops in common use than there are grain crops, as we have shown in Chapters 3 to 5. Second, these forage crops are commonly grown in mixtures which vary widely in composition (Chapter 6), while grain crops are grown in pure stands. Third, the herbage production from forage is influenced much more, in terms of both quality and quantity, by environmental factors (such as the nutrient status of the soil, climate, and management practices) than is the grain from our cereal crops. Most important of all, in the case of our forage crops, we have the option of harvesting at a range of different times. This is not possible in the case of grain crop production.

TABLE 9-5
Proximate feeding stuff analysis (% dry matter) of smooth bromegrass herbage at three growth stages

	Age of herbage in weeks		
	6	8	10
Ether extract	3.5	2.6	2.1
Crude protein	19.2	12.1	6.7
Fiber	19.8	21.6	27.2
Nitrogen-free extractives	45.2	54.7	60.2
Ash	11.6	7.9	5.7
Calcium	0.59	0.61	0.58
Phosphorus	0.41	0.22	0.16
Potassium	3.01	2.75	1.63

(Source: Unpublished data from the University of Alberta forage quality laboratory)

The time of harvesting has a very pronounced effect on forage quality (Table 9-5). As the plant matures, both the ether extractive and crude protein values fall, while the crude fiber and the nitrogen-free extractive values rise. These changes may be explained by studying the development of the grasses (set out in Chapter 1). The major change which takes place in a grass as it ages is the elongation of the stem. The juvenile grass is composed almost entirely of leaves with very short internodes. Leaves contain a high proportion of crude protein and ether extracts, while stems have values for the nitrogen-free extracts' fraction (which contains hemicellulose) and for the crude fiber component (Table 9-6).

Table 9-6 shows the increase with age in the proportion of stem to leaf. It is also evident that, as the plant matures, crude protein and ether extractive values fall, while

TABLE 9-6
Leaf:stem ratio and proximate feeding stuff analysis (% dry matter) for different growth stages of timothy

Sample data	Leaf-stem ratio[a]	Crude protein		Ether extractives		Crude fiber		N-free extractives	
		L[b]	S[b]	L	S	L	S	L	S
20 May	2.57	21.7	14.1	3.8	2.9	19.1	23.5	48.3	49.6
2 June	1.30	17.2	11.4	4.7	2.5	23.8	29.7	47.8	48.3
16 June	0.39	18.5	7.6	4.1	2.6	26.1	32.6	43.3	50.6
30 June	0.35	12.3	4.4	3.3	1.7	26.9	31.6	48.7	52.2
14 July	0.20	11.1	3.4	3.2	1.3	30.6	32.4	46.1	57.9

[a]Ratio by weight
[b]L = leaf; S = stem

(Source: Waite and Sastry, 1949, J. Agric. Sci. 39:174)

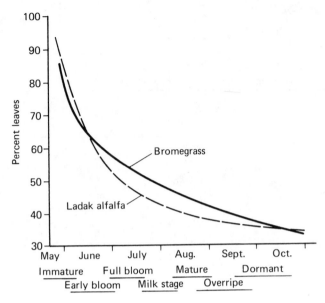

FIGURE 9-3
Percentage of leaves (by weight) at various stages of development for smooth bromegrass and alfalfa. (Source: Canadian Forage Crops Symposium, 1969)

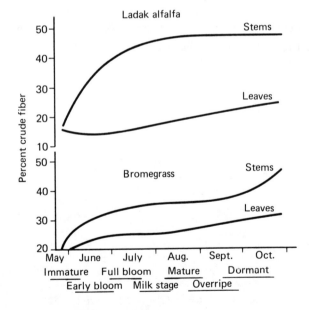

FIGURE 9-4
Proportion (by weight) of crude fiber in alfalfa and smooth bromegrass herbage harvested at various stages in the growing season. (Source: Canadian Forage Crops Symposium, 1969)

the percentage crude fiber increases in *both* stem and leaf. These changes arise primarily from the development of the structural carbohydrate material, consisting mainly of cellulose, hemicellulose, and lignin, which the plant needs for support as its bulk increases. Thus, the changes in proximate feeding-stuff values set out in Table 9-6 represent the combined effect of two factors: an increase in the stem-to-leaf ratio and an increase in structural carbohydrate material as the plant matures.

So far we have discussed only the grasses. Does a similar situation exist for the legumes? Although there is no time of rapid elongation for the legume stem, these species do undergo substantial stem thickening, and marked increases in stem weight take place before flowering. As a result, the percentage of leaves by weight, before and at the time of flowering, is lower for legumes than for grasses (Figure 9-3). The differences discussed are clearly illustrated by Figures 9-4 and 9-5 for percentage of crude protein and crude fiber, respectively. Table 9-7 shows a decline in the crude protein content and digestibility of alfalfa leaves as the plant ages. Alfalfa stems give a much higher crude fiber value than do those of smooth bromegrass during all but the very earliest part of the growing season. The reverse is true for the leaves. Crude protein is higher throughout the season for all parts of alfalfa than it is for the equivalent parts of smooth bromegrass. Thus, the relationship between the leaf-to-stem ratio and plant age is the essential determinant of forage quality.

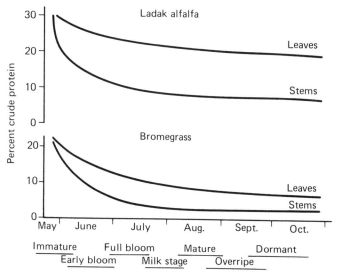

FIGURE 9-5
Proportion (by weight) of crude protein in alfalfa and smooth bromegrass herbage harvested at various stages in the growing season. (Source: Canadian Forage Crops Symposium, 1969)

TABLE 9-7
Effect of age of alfalfa herbage on forage quality in New York State

Herbage	Crude protein (%)		In vitro digestibility (%)	
	Leaf	Stem	Leaf	Stem
2 Weeks old	39.8	32.3	93.6	88.1
6 Weeks old	25.3	9.5	91.5	56.1
10 Weeks old	26.6	9.4	86.8	50.6

(Source: Kaln and Fick, 1981, Crop Sc. **21**)

CLIMATE AND QUALITY

Climate affects quality indirectly by determining the duration of the plant growth stages set out in Figures 9-4 and 9-5. This is especially important for the immature or juvenile stage, when leaves predominate. In fact, plant *age* is a relatively unimportant factor as far as forage quality is concerned. It is the plant's *stage* which determines quality. *Age* is obviously associated with stage; but under different climatic circumstances, plants may move through their growth stages at different rates.

FERTILIZATION AND QUALITY

Fertilizers are normally used to increase forage quantity, but since plant tissue reflects the mineral constituents of the soil in which the plants are grown, quality is also greatly influenced. The herbage is especially responsive to the calcium, phosphorus, potassium, sulfur, and nitrogen content of the soil. Where species are grown in a pure stand, the effect of these minerals on the plant is direct. Where mixtures are involved, it is often the indirect effect that is most important. For example, the application of a nitrogenous fertilizer to a pure stand of any grass results in an increased crude protein value for the herbage. If, however, the same amount of fertilizer were applied to a grass–legume mixture such as brome–alfalfa, there may be no change in the crude protein values for the herbage *as a whole!* The crude protein content of the grass component of the mixture would still increase, as would the growth rate and vigor of the grass plants. This, in turn, would suppress the legumes. Any nitrogen which they take up from the soil would be offset by a reduction in the fixation of atmospheric nitrogen (see Table 13-1), resulting in little change in the total nitrogen available to the plant. Since the crude protein content of the legumes is higher than that of the grass, an increase in the overall crude protein production from the pasture may be small or nonexistent.

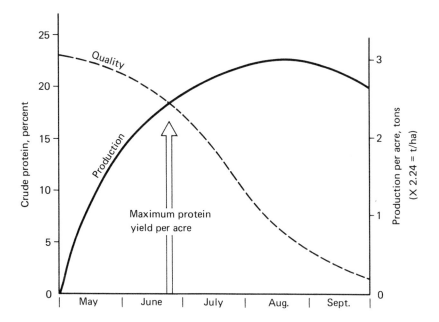

FIGURE 9-6
Relationship between forage yield and forage quality.

INFLUENCE OF MANAGEMENT ON QUALITY

When a forage is grown for hay, the main objective is the production of bulk, but an attempt is made to achieve this while keeping the loss of forage quality to a minimum. Figure 9-6 illustrates well the compromise which the farmer must make in balancing quality against quantity. When forage is grown for use by a grazing animal or where it is grown for drying (dehydration, see Chapter 11), quality is maintained because the management objective is to keep the plant at a juvenile stage by repeated defoliation.

FURTHER READING

Bauengardt, B. R., 1970, Regulation of feed intake and energy tolerance. *In* A. T. Phillipson, ed., Physiology of Digestion and Metabolism in Ruminants, Proc. 3rd Int. Grassland Symp., Cambridge, England, p. 235–53.

Hangate, R. E., 1966, The Rumen and Its Microbes, Academic Press, New York.

Nielsen, K. F., 1969, Canadian Forage Crop Symposium, Western Co-op. Fertilizers Ltd., Calgary, Canada.

Sullivan, J. T., 1966, Studies of the hemicellulose of forage plants, J. Anim. Sci. **25**: 83–86.

Van Soest, P. J., 1967, Development of a comprehensive system of feed analysis and its application to forage, J. Anim. Sci. **26**:119–20.

Van Soest, P. J., 1970, The Chemical Basis for the Nutritive Evaluation of Forage, Proc. Nat. Conf. on Forage Quality Eval. and Util., University of Nebraska, Lincoln, pp. U1–U19.

Worden, A. N., K. C. Sellers, and D. E. Tribe, 1963, Animal Health, Production and Pasture, Longmans, Green & Co. Ltd., London.

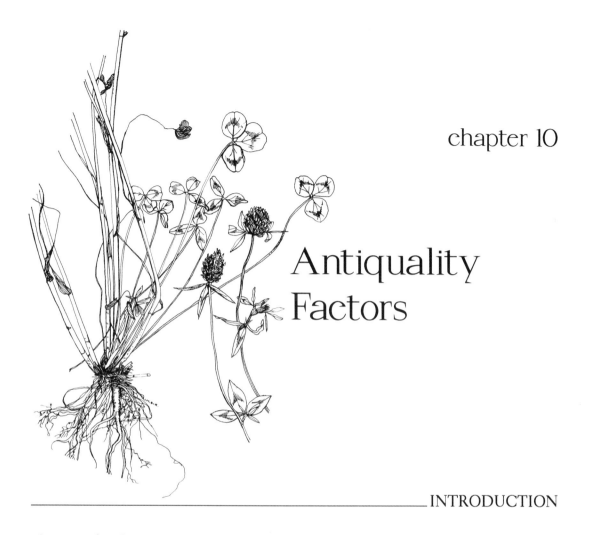

chapter 10

Antiquality Factors

INTRODUCTION

The *antiquality* characteristics of a forage are those factors which contribute to toxicity syndromes, poor animal gains, reproductive difficulties, or low intake under circumstances where the forage rates well when evaluated in the ways described in the previous chapter. In recent years, the combinations of improved livestock breeds, high-producing forage cultivars, and fertilizers have revealed undesirable chemical substances, which previously were not evident, in our forage crops.

The study of antiquality factors in forages falls into the field of allelochemistry, which has been defined as "the stimulation or inhibition of one organism by compounds produced by another." Forage plants produce a wide range of such compounds which act on all types of organisms, from higher animals to insects, fungi, bacteria, and other plant species. No doubt, some of these substances are the outcome of natural selection and form part of the forage plant's survival mechanisms. More often, however, they are evolutionary dead ends which are of no value to the plant. Such substances, the secondary metabolites, are the "side products" of pathways that lead to primary metabolites essential for the synthesis of RNA or DNA, for energy production, for cell structure, or for the buildup of complex protein molecules. The secondary metabolites, which play no role in vital plant processes, are classified into five major groups (Table 10-1). Substances in the major groups will be considered here in relation to the animal disorders which they cause.

TABLE 10-1
Biosynthetic classification of secondary metabolites

Group	Substances included in group
Terpenes	Essential oils, resin acids, rubber and plant pigments (i.e., carotene)
Steroids	Glucosides (including cyanogenetic glucosides), saponins
Phenylpropanes	Coumarin
Acetogenins	Flavonoids and tannins
Alkaloids	Teratogenic alkaloids

(Source: Anti-quality Components of Forages, CSSA Special Publication 4, December 1973. By permission of the Crop Science Society of America)

CYANOGENETIC GLUCOSIDES

These substances, which are harmless in themselves, may be broken down to produce hydrocyanic acid (HCN), sometimes called prussic acid. Concentrations of more than 100 mg of hydrocyanic acid per 100 g of dry plant tissue are considered dangerous. HCN causes death by asphyxiation at the cellular level. It is readily absorbed into the blood and is carried throughout the animal's body, forming an inactive complex with cytochrome oxidase, an enzyme essential for cellular respiration.

HCN may be released from the cyanogenetic glucosides in two ways: either by the action of the rumen microflora or by plant enzymes liberated when plant tissue is injured as a result of cutting, grazing, wilting, or freezing. Under these circumstances, the enzyme β-glucosidase will remove the sugar component from the cyanogenetic glucoside molecule, and a second enzyme, oxynitrilase, will then convert the remainder of the molecule to aglycone and hydrogen cyanide (HCN). Levels of cyanogenetic glucosides are high in young plant tissue when high applications of nitrogen fertilizer are used and also under environmental conditions which cause plant stress. Fortunately, not all forage species contain potentially dangerous amounts of cyanogenetic glucosides. High levels are commonly found in sorghum, sudangrass, and johnsongrass. Sorghum hybrids frequently have levels lower than the parent plant. In the legumes, these substances are present in white clover and in birdsfoot trefoil. The presence of cyanogenetic glucoside in both species is under the control of a single gene, so it would be relatively simple to breed varieties free from these substances.

SAPONINS

These substances, like the glucosides, form part of the steroid group (Table 10-1). Producing a soaplike foam in water, they cause hemolysis of red blood cells, are toxic to cold-blooded animals, and depress the growth of chicks and the egg production of hens. As well, they are either toxic to or will suppress the growth of some plant species (e.g., suppression of the germination of cotton seed). They are, in fact, antimetabolites capable of a wide and diverse range of biological activity.

Alfalfa cultivars show a marked response to selection for high saponin content, but the selection response for low saponin content is small. The saponins in alfalfa, known as triterpenoid saponins, are present in all parts of the alfalfa plant. Low and high saponin lines have been tested for resistance to attack by a number of insects and diseases. Only in the case of the pea aphid (*Acythosiphon pisum* Haris) could it be shown that resistance is decreased where saponin content is low.

BLOAT

Plant saponins in alfalfa have been studied widely because they were believed to be associated with *bloat*. The microbial fermentation of forage material in the rumen leads to the production of a large amount of gas which is then absorbed into the rumen wall, passed into the omasum or, most commonly, expelled by eructation (belching). The eructation of gas is prevented by the development of a stable viscous foam in the rumen. This foam may hold the gases in large pockets (free gas bloat) or small pockets (foamy bloat), and it may be caused by feeding a high-concentrate ration in the feedlot, by feeding legume hay, or, most commonly, by grazing pastures which contain a large portion of leguminous material.

Failure to expel the gases held in the foam results in the rumen becoming distended so that it either presses against the diaphragm, immobilizing it, or constricts the main dorsal vessels through which blood passes to and from the heart. In either case, animal death is due to asphyxiation. Bloat is the cause of substantial losses to cattlemen throughout the world. In New Zealand, some 17,000 to 20,000 dairy cattle die from bloat each year. In the United States, it is estimated that farmers lose livestock valued at about $80 million yearly, while in Canada, annual losses are in the region of $22 million. Consequently, means of preventing bloat have been widely studied. The most common approach has been to try to prevent or reduce the development of foam in the rumen by reducing, by plant breeding methods, the amount of foam-causing substances in the plant material.

The saponins, which are capable of increasing surface tensions to levels that can withstand the gas pressures which develop in the rumen, were the first substances to be considered as the main causal agents of bloat. While they do cause increased and irregular respiration, which is frequently found in bloating animals, it has proved difficult to associate saponin levels with incidents of bloat.

MacArther and Mutlimore, in 1966, working in British Columbia, presented evidence to show that another important foaming agent was a plant protein known as 18S. They showed that incidents of bloat were associated with high 18S levels in a range of legume species. Alfalfa contains 4.5% to 5.2% 18S protein. Other legumes known to cause bloat (red clover, white clover, sweetclover, and alsike clover) contain similar amounts. Legumes which do not cause bloat (birdsfoot trefoil and sainfoin) and the grasses contain less than 1% 18S protein. This protein substance, which is so named because its sedimentation coefficient is 18 Svedberg units, is also known as the fraction I protein. It has a molecular weight of 500,000 and, since it is one of the cytoplasmic proteins is readily digestible. Also, any reduction in fraction I protein would seriously impede photosynthetic rates, since it consists of ribulose 1,5-diphosphate carboxylase, the main carboxylating enzyme in the C_3 pathway. The heritability of this trait is high, and selection would rapidly decrease 18S protein levels; however, both quality and quantity of forage production would decline if 18S protein levels were reduced by plant breeding methods.

Studies of 18S protein indicate that stable foam found in the rumen results from protein molecules, which are spherical when in solution, reaching the surface of the rumen fluid without being broken down by microbial action. Under these circumstances, the molecules uncoil, become insoluble, and are then capable of stabilizing the foam. Studies in New Zealand show that both fraction I and fraction II proteins (a mixture of proteins with molecular weights between 10,000 and 200,000) act in the same way to stabilize foam. Recent work has shown that certain tannins are capable of precipitating the proteins found in bloat-causing foam. The nonbloating legumes contain large amounts of these tannin substances. While the plants are alive, the tannins are held in vacuoles so as to prevent precipitation of the plant's own protein. The world collection of the genus *Medicago*, which is available at the University of Alberta, has been surveyed for tannin content. None of the *Medicago* species, nor any

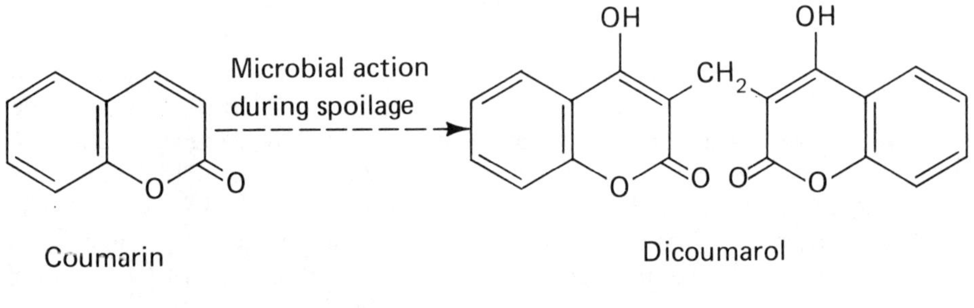

FIGURE 10-1
Conversion of coumarin to dicoumarol. (Reproduced from Anti-quality Components of Forages, CSSA Special Publication 4, December 1973. By permission of the Crop Science Society of America)

of the cultivars in common use, have been found to have high tannin levels. Consequently, it would seem that for the present, management, rather than plant-breeding solutions, must be used to combat bloat.

Bloat prevention may be achieved by a combination of pasture management and the use of antifoaming agents. The proportion of bloat-causing legumes in a pasture should be 50% or less. Before turning an animal into a pasture containing a high proportion of a bloat-causing legume (over 30%), it is wise to feed a grass, hay, or forage containing tannins (e.g., sudangrass).

The antifoaming agent most commonly used is poloxalene (a polyoxypropylene-polyoxyethylene block polymer), which is sold under the name Bloatguard. It is available in molasses "licks" or may be mixed into the concentrate part of the feed. Antibiotics will also prevent bloat, as will a number of natural oils, such as soybean oil, corn oil, peanut oil, or olive oil. However, these substances are rapidly degraded in the rumen and consequently call for large and frequent doses, which are costly to administer.

In Australia, antibloat capsules are available. These are 15-cm gelatin cylinders, 4 cm in diameter, which split down the middle and are hinged on one side. The capsule, which splits open in the rumen, becomes too large to be regurgitated, and releases a foam dispersing detergent at the rate of 6 g/day for 24 days.

SWEETCLOVER BLEEDING DISEASE OF CATTLE

Both internal and external bleeding may occur in livestock as a result of the presence in their blood of the anticoagulant dicoumarol. Dicoumarol is produced from coumarin, a harmless substance found in sweetclover, which breaks down if overheating or spoilage takes place during hay or silage making (Figure 10-1). Coumarin, in turn, is not normally present in the living sweetclover plant, but is produced as a result of freezing, drying, or, most usually, maceration (as is hydrogen cyanide in other species). Coumarin has a sweet smell, which will be familiar to anyone acquainted with new-mown sweetclover hay. However, it has bitter flavor. This may well confer on sweetclover an advantage which has been developed through many generations of natural selection. Insects dislike the bitter taste of coumarin and reject sweetclover in favor of other plants. Unfortunately, livestock palatability is reduced by this substance, as well.

Spoilage is very likely to occur in making sweetclover hay, since the plant's leaves are small and fine and dry quickly, while the stems are thick and retain moisture. In an attempt to retain a high proportion of leaves, sweetclover hay is often stacked before the stems are fully dry. Chopping, which is a feature of many modern hay-making systems (see Chapter 11) and a common practice in silage making helps to overcome this problem.

As in the case of plants containing cyanogenetic compounds, the genetic control of the presence of plant substances from which dicoumarol might be produced is simple. It has been possible to breed cultivars in which dicoumarol production is negligible. Such cultivars are, however, low yielding.

TANNINS

Tannins are acetogenins, the fourth group in the biosynthetic classification of secondary metabolites (Table 10-1). They are polymeric phenolic compounds which differ from other polyphenolic compounds by having strong protein-binding properties. In many forages, they are responsible for a bitter taste, which has been shown to result in a reduction in animal intake. Tannins also reduce digestibility. This is believed to result from the inhibition of the cellulolytic and pectinolytic enzymes.

The protein-binding characteristics of the tannins protect the protein part of the ration by preventing bacterial deamination in the rumen. The protein may then be effectively absorbed in subsequent parts of the animal's digestive system. Thus, the addition of tannins to ruminant rations may have a beneficial effect by increasing nitrogen utilization. This practice has been shown to increase the weight gains of young lambs. Such gains, however, are often outweighed by a decrease in intake and in digestibility.

FLAVONOIDS

Flavonoids, which cause reproductive failures in livestock, are placed with tannins in the fourth group of secondary metabolites (Table 10-1). They were first detected in 1944 in a flock of sheep grazing on subterranean clover. The flavonoid coumestrol, which was subsequently isolated from alfalfa, proved to be 30 to 40 times more potent than the isoflavone estrogens first detected in subterranean clover. Coumestrol is, however, much less active than the natural estrogenic substances which livestock produce or than synthetic diethylstilbestrol (DES).

In forage plants, estrogenic flavonoids may build up to physiologically active concentrations when the plants are diseased. Leafspot fungi are the most common cause of such accumulations. The proportion of estrogenic flavonoids in a forage plant depends, in fact, on a variety of environmental factors, which include season, temperature, stage of growth, and degree of defoliation.

Flavonoids are not entirely detrimental. Clover has been shown to increase the growth rate of lambs. This observation is attributed to the estrogenic activity of the plant material. The ban on the oral use of DES, which increases the efficient use of feed by 10%, but which has been shown to produce tumors in mice, has led to increased interest in obtaining liveweight gains from naturally occurring estrogens. So far, plant selection has not produced strains with high enough concentrations of estrogenic substances to be useful.

It is difficult to assess the magnitude of either the beneficial or the detrimental effects of flavonoids. Sheep, for example, will sometimes drop their lambs prematurely or fail to conceive. Such a moderate suppression of fertility in livestock may remain undetected and consequently would not be associated with pasture condition. Also, the infertility itself may be transient.

ALKALOIDS

Alkaloids, which number over 2,000, are found in 10% to 15% of all vascular plants. They form a rather heterogeneous group, but are similar in that they act chemically as bases, contain nitrogen, are of plant origin, are complex, and are pharmacologically active. The most fully studied plant alkaloids found in forages are the eight compounds present in reed canarygrass (Figure 10-2). This grass produces very high yields where soil moisture is good, but its value is greatly reduced when the alkaloid content is high. For example, if an organic solvent extract of reed canarygrass is sprayed onto low-

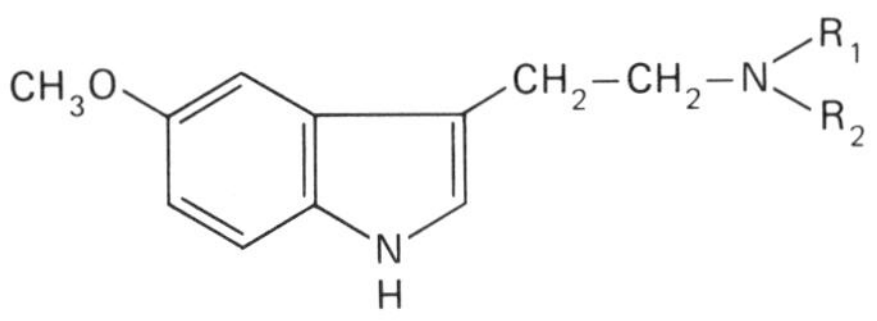

Ia $R_1 = CH_3, R_2 = H$
Ib $R_1 . R_2 = CH_3$

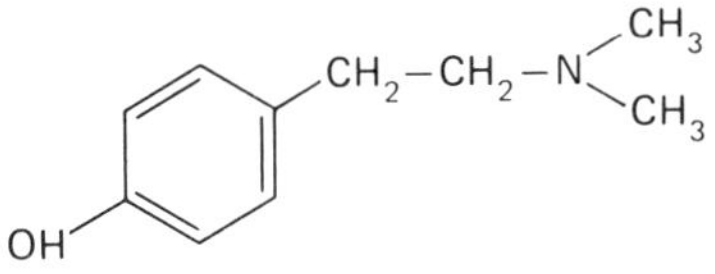

II

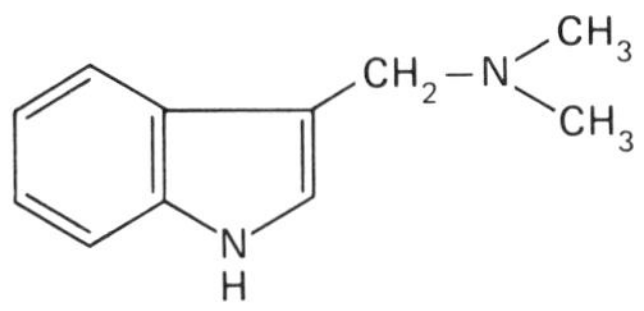

III

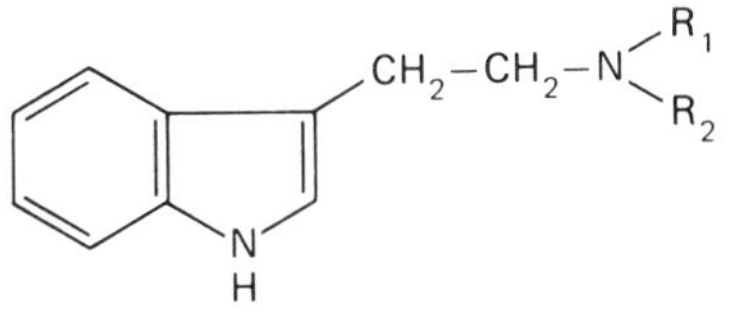

IVa $R_1 = CH_3, R_2 = H$
IVb $R_1 . R_2 = CH_3$

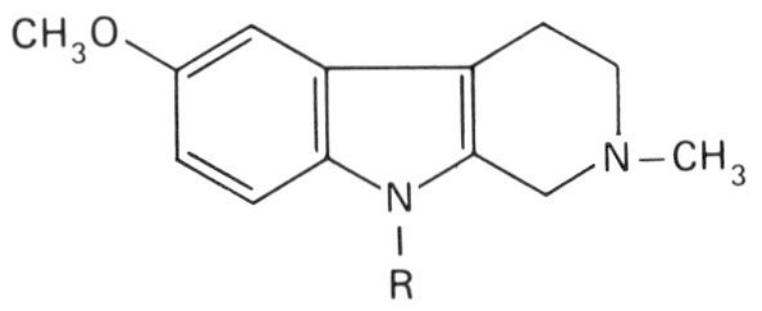

Va $R = CH_3$
Vb $R = H$

(Ia) 5-methoxy-*N*-methytryptamine
(Ib) 5-MeO-*N,N*-dimethyltryptamine
(II) hordenine
(III) gramine
(IVa) *N*-monomethyltryptamine
(IVb) *N,N*-dimethyltryptamine
(Va) 2,9-dimethyl-6-methoxy-1,2,3,4-tetrahydro-β-carboline
(Vb) 2-methyl-6-methoxy-tetrahydro-β-carboline

FIGURE 10-2
Structure of eight alkaloids found in reed canarygrass. (Source: Anti-quality Components of Forages, CSSA Special Publication 4, December, 1973. By permission of the Crop Science Society of America)

alkaloid material, cattle intake, palatability, and digestibility are all decreased. Also, following extraction, the residue of the unpalatable material is readily consumed.

Normally, palatability, intake, and digestibility of reed canarygrass strains with a high-alkaloid content are low, while cultivars with a low-alkaloid content rank high for these same characters. Work with sheep in Australia has shown that the intake of high-alkaloid reed canarygrass was 0.44 kg (about 1 lb) of digestible organic matter per animal per day, while the intake of low alkaloid material was 0.68 kg (about 1.5 lb) per animal per day. No doubt, the intake is reduced in part because the palatability is low. Digestibility, which also influences intake, is reduced because the alkaloid substances interfere with the activity of the microorganisms in the rumen fluid. In the case of *gramine* (a primary alkaloid), dry-matter concentrations of 0.2%, 1%, and 2% gave digestibility values for reed canarygrass hay of 60.5%, 45.8%, and 38.0%, respectively.

Studies show that the alkaloid concentration in reed canarygrass is under genetic control. This trait is highly heritable, and the genetic variance is largely additive. The environment also plays a part in the expression of alkaloid concentration. It is found to increase with high soil fertility, especially where high nitrogen levels are present. Even more marked is the increase in alkaloid concentration which takes place when plants are placed under moisture stress, especially when high temperatures are involved. Also, as the plants age, alkaloid concentrations increase.

There are two tryptamine alkaloids found in reed canarygrass. Both have been linked with a disease of sheep known as *phalaris staggers*. This is a disorder of the central nervous system, affecting both the brain and heart, which may cause animals to suddenly collapse and die. Cattle may also be affected, but deaths are much less frequent.

TALL FESCUE TOXICITY

The symptoms of a disorder called *fescue foot*, which is associated with cattle grazing tall fescue pastures, include loss of weight, dull, rough hair, fat necrosis, lameness, and dry gangrene of the feet, tail, and ears. The back is frequently arched. The animals have elevated temperatures, respiration, and pulse rates. The hind foot has a characteristic red line and loss of hair at the coronary band. This condition occurs most frequently when cattle are grazing tall fescue pastures in the winter months. Usually only part of the herd (about 20%) shows the clinical signs. These animals frequently recover a few days after being removed from the toxic pasture. It would seem that animal variation exists, some individuals being more susceptible than others.

There are many theories of the cause of fescue foot in cattle. Tall fescue contains two types of alkaloids: perloline and perlolidine. These substances, when isolated and given to cattle orally, intraruminally, or intravenously, do not always produce the symptoms of fescue foot. Frequently, they are only mildly toxic.

The vasoconstrictor alkaloids produced by ergots have been considered as possible causal substances of fescue foot. These substances are various indole compounds which are formed in the sclerotia of the fungi which produce ergots and which are known to cause gangrenous ergotism. The same substances have been isolated, in very small amounts, from disease-free tall fescue. Certainly, the clinical symptoms of fescue foot and gangrenous ergotism are similar. However, the amounts of these substances extracted from tall fescue are too small to cause the disease.

A third possibility is that fungi produce mycotoxins or cause the plant to produce and accumulate toxic phytoalexins which, when digested, cause fescue foot. Another theory suggests that the toxic substance is produced in the rumen. The fungus *Aspergillus terreus* is associated with the formation of antibiotics which will kill beneficial bacteria in the rumen. If these beneficial bacteria are not present to break down toxic alkaloids, the animals will be exposed to these compounds. Which of these theories provides the true explanation of the cause of fescue foot remains to be determined. The mechanism by which the toxins (whichever they may be) function inside the animal to produce the clinical symptoms is also unknown.

NITRATE POISONING

Nitrate poisoning occurs when animals eat forage material with a high nitrate content (in excess of 0.35% to 0.45% nitrate in the diet). Under these circumstances, the nitrate is converted to nitrite in the rumen. The nitrite is absorbed into the blood, converting hemoglobin into methemoglobin, a substance which is incapable of transporting oxygen. When it is not fatal, nitrate poisoning produces subclinical conditions, which result in poor animal performance and a general lack of condition. The animal's response to nitrate poisoning is influenced by other components of the ration, particularly the availability of carbohydrates.

The plant produces or accumulates nitrates because the first step of protein synthesis involves the use of these substances. Consequently, anything which influences the sink-source relationship between protein production and nitrate accumulation will influence the nitrate content of the plant's tissue. The most common causes of high nitrate content in forage tissue are the following:

1. High applications of nitrogen fertilizers or high soil fertility.
2. Drought conditions.
3. Damage to plant tissue (such as defoliation as a result of grazing or hail damage), which will stop or reduce photosynthetic activity.
4. Low light intensity.
5. Plant species (some plants convert amino acids to proteins rather slowly).
6. Management (if animals are made to graze closely, they will eat more of the lower stem tissue).

The frequent use of nitrogen fertilizers in recent years has resulted in an increased incidence of nitrate poisoning. The belief that nitrate poisoning occurs only when annual forages are fed is quite untrue. Perennial forages are just as likely to accumulate nitrates where high fertilizer dressings are used.

GRASS TETANY

Grass tetany, or hypomagnesemia, is caused by low levels of magnesium in the animal's blood. Even when magnesium levels are adequate in the herbage, absorption may be low. Sheep and goats are less susceptible than cattle. Pregnant animals are especially prone to this condition. Herbage with a $K+:(Ca+$ and $Mg+)$ cation ratio exceeding 2.2 places animals at greater risk than when cation ratios are low. The cation ratio may be influenced by fertilizer treatments or by soil fertility. For some species like Kentucky bluegrass, crested wheatgrass, tall wheatgrass, and meadow foxtail, the cation ratio is normally lower than for other forages. Incidents of grass tetany are high for grazing ruminants when temperatures fluctuate.

FURTHER READING

Arnold, G. W., and J. L. Hill, 1972, Chemical factors affecting selection of food plants by ruminants. *In* J. B. Harborne, ed., Phytochemical Ecology, Academic Press, London, pp. 72–101.

Buckner, R. C., 1979, Tall Fescue, American Society of Agronomy, Agron. Series No. 20, Madison, Wisconsin.

Cape, W. A., and J. C. Burns, 1971, Relationships between tannin levels and nutritive value of sericea, Crop Sci. 11:231–233.

Matches, A. G., ed., 1973, Anti-quality components of forage, Crop Science Society of America, Special Publication No. 4., Madison, Wisconsin.

Whittaker, R. H., and P. P. Feeny, 1971, Allelochemics: chemical interactions between species, Science **171**:757–770.

Worden, A. M., K. C. Sellers, and D. E. Tribe, 1963, Animal Health, Production and Pasture, Longmans, Green & Co. Ltd., London.

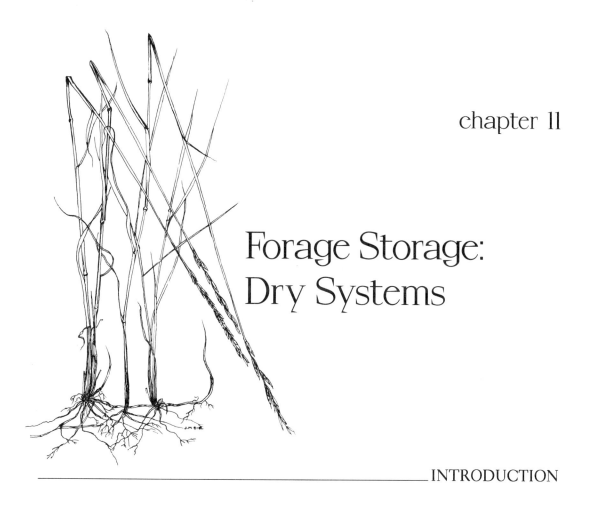

chapter 11

Forage Storage: Dry Systems

INTRODUCTION

Storage, in the widest sense of the word, is one of the main objectives of all plant production. All seeds, including cereal and legume grains, are of value because they represent food material, in a concentrated dry form, which is capable of being stored, transported, and used when and where it is required. Oil seeds, drugs, spices, beverage crops, nuts, and dried fruits, together with cereals of all kinds, are examples of plant products of this type.

The primary reason for the production of many of these crops is to meet our food needs; only the by-products are used for animal production (e.g., crushed oil seeds). During the last quarter-century, however, grain supplies have been so plentiful and prices so low that cereal grains have become a substantial part of livestock rations. Present world prices for human food suggest that this is most unlikely to continue. If it is to remain profitable, animal production will have to depend far more heavily on stored forage products. Forage storage is quite unlike the storage of grain crops, where natural plant processes bring about the only changes needed in the plant material. The moisture content of herbage is high, and the problems of preserving such material are many and complex.

Green herbage continues to "live" after it has been cut. On a bright sunny day, such material photosynthesizes, produces sugars, and increases in dry weight for several hours. Eventually, depending on the initial water content of the herbage, the plant cells lose their rigidity and the crop dies. "Sugars" in the plant juices then start to oxidize and proteins start to break down. When the plant is alive, it is able to resist attacks by the many bacteria and fungi which are always present on its outer surface. Organisms of this type are, however, well able to decompose dead or dying tissue. The aim of forage storage is to stop such destruction and to preserve both the quantity (yield) and the quality (feeding value) of the herbage.

There are two ways in which these objectives may be achieved. The crop may be dried, by hay making or dehydration, or it may be preserved by pickling or *ensiling*. Drying reduces the moisture content to a point where both chemical breakdown and microbial action cease. Ensilage achieves the same results by reducing the pH. That the losses which occur before the preservation process is complete be kept as small as possible is an essential feature of either system. In this chapter, we will consider the systems which depend on desiccation, leaving ensilage for the next chapter.

HAY-MAKING PRINCIPLES

The origin of hay making is uncertain, but it is believed to have developed very early in our history. The Roman writer Columella described the process about 2,500 years ago. From that time, hay-making practices remained virtually unchanged until toward the end of the nineteenth century. The hay loader (1874) and the side delivery rake (1893) represent the first steps taken away from the scythes and pitchforks of antiquity. Hay making today is still demanding in terms of energy, time, and human effort. Quite recently, a great many new hay-making systems have been evolved. However, this still remains an important area of challenge and one in which production costs could be decreased by improved systems.

What, then, is the object of hay making? We seek to produce, with minimum dry-matter loss and minimum expenditure, a stable animal feed of good nutritive value. In practice, the crop in the field contains from 90% moisture by weight for a young, immature crop down to 75% moisture for mature, more fibrous herbage. The hay is handled in such a way that its moisture is reduced from these levels to about 25% moisture by sun and wind as the crop lies in a *swath* or *windrow*. A *swath* is the cut forage lying in the field, while a *windrow* is a row of forage formed by raking a swath together or formed directly by a mower-conditioner or windrower (see Figure 11-1). When the crop is gathered up from the windrow or swath, further drying in the barn, the stack, or the field may be needed.

If yields of hay are good, then, during the whole process, a total of between 3 and 4 tonnes of water/ha (1.35 to 1.8 tons/acre) must be evaporated. Three-quarters of this can be removed on the day the grass is cut, if both weather and management are favorable. To achieve this high, early drying rate, windrows should not be tightly packed and should be held off the ground by the stubble (75 to 100 mm or 3 to 4 in. high) left when the crop is cut. The air will then be able to circulate under the windrow, increasing the rate of drying. Long stubble length is also advantageous because the

FIGURE 11-1
Windrowing a crop of alfalfa. (Photograph courtesy of Deere & Co.)

regrowth shoots of a legume crop like alfalfa come from buds on the lower part of the stem. A high cut ensures that an adequate number of buds are present for rapid regrowth after defoliation, thereby keeping the pasture productive. If the plant is cut near to ground level, leaving few stem buds, crown buds will develop. Such buds grow more slowly than stem buds, thus slowing production rates.

The size and compaction of a windrow is important in relation to the climatic conditions which exist when the hay is being made. For example, in addition to the moisture present in the herbage when it is cut, water will also be formed by the oxidation of plant "sugars." A tightly packed windrow may, then, become *wetter* following cutting. Also, surface drying may result in large differences in water content between the top and center of a tightly packed windrow. Under these circumstances, leaves on the outside become brittle and are shed when the crop is handled. There are also losses of up to 10% dry matter due to respiration in the moist, center part of the windrow. Consequently, temperature, number of sunshine hours, and the prospect of rain at hay-making time are all important in determining harvesting methods and equipment. The rate and the total amount of drying which might be expected in a swath or windrow in a particular environment should match the harvesting method used.

Whatever the type of windrow selected, the diffusion of water vapor slows as the forage's moisture content falls and reaches about 30%. At this point, the thick plant stems retain more moisture than do the leaves. Uneven drying will result in a dry-matter loss which will include many leaves, nutritionally the most valuable part of the plant.

This problem is overcome by mechanically *conditioning* the herbage. This is done, at the time when the crop is cut, by passing it between two rollers which are attached to the mower or windrower or are part of a separate machine. These rollers are of two kinds: either there are two corrugated rollers which mesh like gears [called a *crimper;* see Figure 11-2(A)] or one roller is smooth and the other is ribbed [called a *crusher;* see Figure 11-2(B)].

Figure 11-3 compares conditioned and unconditioned alfalfa stems. The soft, moist tissue in the center of the unconditioned stem dries at about the same rate as the leaf, once the stem is cracked (conditioned). The use of a conditioner increases the drying rate and reduces the time when the crop is vulnerable to weather damage. This, in turn, prevents leaf loss; consequently, the quality and palatability of conditioned hay are superior to unconditioned material. Conditioning is especially important for sweet-clover, which has very fine leaves and thick stems. It is also important for sudangrass, which has a woody stem covered with a waxy cuticle that will hold moisture for long periods.

To maintain quality, it is desirable to remove hay from the windrow as soon as possible. Exposure to rain leaches soluble nutrients (which are also the most digestible) from the swath, while exposure to the sun denatures the carotene in the herbage. Also, damp, moist conditions prolong respiration and so increase nutrient losses. In total, 30% of the dry matter is frequently lost in field operations. However, if the hay is removed from the field with a high moisture content (30% or over), losses will occur in storage due to the continuing respiration of living plant material.

Storage losses due to fermentation, which is possible only when moisture levels are high, are accompanied by a rise in temperature. At 32°C (90°F), respiration is greatly reduced; at about 45°C (113°F), the plant cells die. At this stage, bacteria and molds become active, further increasing the rise in temperature. At 70°C (150°F), chemical oxidation commences and, should enough air be present, spontaneous combustion

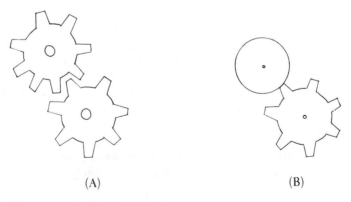

FIGURE 11-2
A side view of (A) crimper rollers and (B) crusher rollers. Wide, cleated rubber rollers with some of the features of both types are also available.

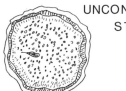 UNCONDITIONED STEM

 CONDITIONED STEM

FIGURE 11-3
Comparison of a conditioned and an unconditioned stem.

results. At this point, loss of the preserved material occurs. Although this does not happen frequently, respiration losses alone can account for the reduction of the total available nutrients by as much as 40%.

Timeliness of harvesting is of the utmost importance in obtaining the maximum yield of nutrients. The inverse relationship between the yield and the quality characteristics of a forage were discussed in Chapter 9 (see Figure 9-6). This relationship determines the optimum harvest period of a hay crop under any specific environmental circumstance. Figure 11-4 shows that, while dry matter continues to increase throughout the life of the plant, the crude protein content decreases as the plant matures. High-capacity equipment (or additional equipment) may be needed to complete harvesting within the optimum time period.

HAY-MAKING METHODS

National and international machinery companies have made available to the farmer hay-making systems which are many and varied. These systems are best considered in relation to the kind of hay they produce. These include long, chopped, shredded, baled, and wafered or pelleted hay.

Long Hay

Long hay was the earliest form made. It was the kind of forage stored in ancient Rome and it is still produced today. The crop is cut into a swath (Figure 11-5), gathered into a windrow, and then loaded and transported for storage. Frequently, these operations are not combined, each being carried out by a different machine. This crop, which is not chopped at all (hence the name "long" hay), is bulky and open in structure, permitting air movement. Consequently, it may be safely stored with moisture content of 25%.

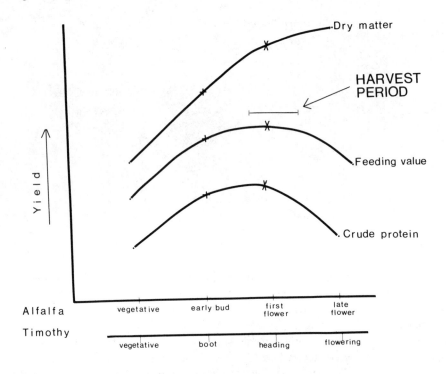

FIGURE 11-4
Forage production measured by "feed value," rather than bulk alone, shows the importance of timeliness and speed in harvesting. (Source: Canadian Forage Crop Symposium, 1969)

For long hay, the herbage is cut using a sickle-bar mower, then gathered by a dump rake, side delivery rake, hay loader or buck rake, and loaded into a hay wagon or field stacked. The object of raking is to move the mowed hay into a fluffy windrow (see Figure 11-6), with the green leaves on the inside. There they are protected from the sun, and will dry slowly, so as not to become brittle before the stems are dry. A rake may also be used to turn a rain-soaked windrow onto dry ground to achieve fast, uniform drying.

Whatever the type of hay produced (long, chopped, or shredded), the preparation of the windrow is important. While it is possible to mow, condition (within 30 min of mowing), and subsequently collect into a windrow in separate operations, all three can be accomplished in one pass by using a mower-crusher-windrower. This machine allows drying to take place in the least possible time. This is most important to reduce leaf and nutrient losses and to lower the risk of losses due to bad weather.

Figure 11-7 shows hay-drying curves for herbage harvested by three different methods. Curve 1 summarizes data for herbage placed in a heavy windrow and left

FIGURE 11-5
This mower-conditioner is being used to speed drying and so obtain better-quality hay. (Photograph courtesy of Sperry-New Holland)

FIGURE 11-6
This open windrow has dried to a moisture content of 25% or less and is ready for baling. (Photograph courtesy of Sperry-New Holland)

202 / Forage Storage: Dry Systems

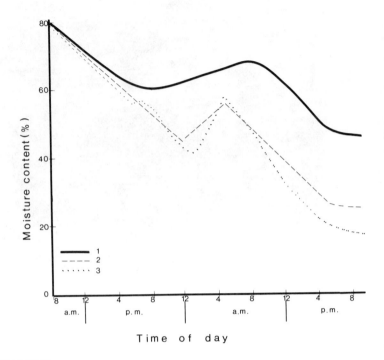

FIGURE 11-7
Hay drying curves for (1) a heavy windrow; (2) swathed; (3) multitreatment unit. (Source: Drawn from Agriculture Canada information)

to dry. It indicates that this treatment gives a slower drying rate than when the material is mowed and left in a swath (curve 2). Dependent on the weather, only a multitreatment unit (curve 3) produces dry herbage suitable for baling by the end of the second drying day. The moisture content of the herbage often increases in the early morning hours because of the dew (Figure 11-7). It should be noted that the treatments which result in the greatest moisture loss also take up the most moisture in the mornings.

Chopped Hay

One of the advantages of chopping hay is that its physical properties are greatly improved, since it will flow freely. Also, it has a *smaller* storage volume and the animal intake is higher than for long hay (see Chapter 9 and Figure 9-2). There is, however, one important disadvantage. Since the material has a higher density, air movement is reduced, so the safe moisture content for storage is lower than that for long hay. The appropriate moisture percentage will vary with the size to which the hay is chopped.

Achieving these low-moisture contents may result in leaf loss due to shattering and increased field loss due to weather conditions. When chopped hay is produced, it is essential that the hay material be conditioned.

The chopped hay is mowed by a sickle-bar mower, conditioned by crushers or crimpers, and then windrowed using a side delivery or finger wheel rake. Self-propelled windrowers or flare cutters, both fitted with conditioners, may also be used. In all cases, the hay is left in a windrow to dry. A forage harvester is used to gather the hay from the windrow and chop it. Suitable wagons haul the chopped hay from the field and into the store. The chopped hay may be moved from the store by automated bunk feeders. The overall system is similar to that which is frequently used with green material for silage making.

Shredded Hay

Hay of this type is lacerated along the length of the stem rather than being cut across the stem. The windrow is prepared in the same way as it is for chopped hay. A cutter-blower can be used to pick up the windrow, lacerate it, and fill a wagon. Numerous machinery manufacturers now have equipment available which combines some laceration with some chopping of the hay.

Baled Hay

In spite of the fact that it is difficult to mechanize bale handling, the bale is the most widely used method for making hay. The automatic pickup baler came into general use in about 1940. The advantage of this equipment is in its ability to compress forage into a dense unit which can be stored and transported at a reduced cost (Figure 11-6), making hay feasible as a cash crop. Baling is, in fact, simply a way of packaging hay. The traditional bale weighs about 24 kg (53 lb), has a density of about 160 kg/m^3 (270 lb/yd^3) and is made from herbage with a moisture content of between 20% and 30%. In recent years, a number of mechanized systems have become available which aim to reduce the labor involved in handling these conventional bales. Bale chutes, which move bales directly into a trailing wagon, are available. A bale "thrower" fulfills a similar function. Bales may also be dropped in groups by using a bale binder, which forms the bales into a pyramid of 3 to 15 units, thus providing some weather protection. Bales may also be grouped to form "pallets." The bales are then assembled by an accumulator and loaded as a unit by using a grapple attachment on a front-end loader. Automatic bale wagons will also pick up and drop individual bales, as shown in Figures 11-8 and 11-9.

Large Bales

A tractor with a front-end loader is able to carry large, single, stable units of over 1 tonne. This has led to the production of very large round or rectangular bales. Because the equipment required to produce the large rectangular bales is expensive, the simpler

FIGURE 11-8
Bales are picked up in the field and formed into a stack. (Photograph courtesy of Sperry-New Holland)

FIGURE 11-9
An automatic bale wagon dropping bales one at a time along the feed line. (Photograph courtesy of Sperry-New Holland)

round bale equipment has proved to be more popular (Figure 11-10). A round baler consists of a simple set of countermoving belts which wrap windrowed hay into a solid cylinder (Figure 11-11).

There are two types of round balers. One *picks up* the hay and rolls it into a large cylinder, and the other rolls the hay *along the ground.* The main advantage of the large round bale is its ability to shed water and resist weathering more effectively than the conventional bale. Round bales, which usually weigh between 460 and 1,500 lb. Depending on the equipment used, the harvesting rate will vary from 5.5 to 13.5 T (5 to 12 tons) per hour.

The baling losses from round balers are very similar to those with the smaller conventional balers (3% to 15%), but field losses are higher when harvest rates are low due to light windrows. Low moisture content (10%) and overmature or badly weathered hay also give high baling losses, since it is difficult to handle the round baler in such a way that it will pick up this type of material. It may, in fact, in some areas, help to wait for dew or light rain.

Operator experience is an important factor with round baler equipment. The pickup is as wide as the bale chamber into which hay must be fed evenly to produce a uniform bale. This problem could be overcome if manufacturers made the pickup wider than the bale chamber. Tractor wheels should be set well apart to avoid running over the edges of the windrows.

After baling, the bales should be removed from the field as soon as possible to prevent damage to the plants located under the bale [Figures 11-12 and 11-13(A) and

FIGURE 11-10
Round baler picking up hay from a windrow.
(Photograph courtesy Sperry-New Holland)

FIGURE 11-11
Round bales. (Photograph courtesy of Sperry-New Holland)

FIGURE 11-12
Round bales should be moved from the field as soon as possible. (Photograph courtesy of Sperry-New Holland)

(A)

(B)

FIGURE 11-13
Round bales being made ready for transport.
(Photograph courtesy of Sperry-New Holland)

(B)]. In the storage area, the bales should be placed, singly, with a space between individual bales, and with the same side on the ground as when they were dropped from the baler. These arrangements will provide the maximum protection against rain, which can penetrate right through the bales if they touch at the sides.

The giant square baler produces a bale about 1.5 m wide, 1.5 m high, and 2.3 to 2.5 m long (5 by 5 by 8 ft). The bales weigh about 850 kg (1,800 lb) when made from alfalfa hay and 350 kg (770 lb) when the material baled is barley straw with a moisture content of 10%. With a square baler, the harvesting rate under good conditions is 13 to 18 T (12 to 18 tons) per hour. Baling losses are similar to those for conventional bales. The bottom, sides, and front ends of these types of bale are denser than the top and back end. Consequently, if the bales are stacked, they should be placed with the front ends above one another. Or if the bales are turned on their sides, the bottoms should be placed above one another. Improperly made stacks will fall over.

Barn Drying

When weather conditions are *not* favorable, the time of removal of hay from the windrow may become a compromise between accepting a moisture content too high for safe storage and leaving the crop in the field to suffer further leaching and physical loss. The solution to this dilemma is some form of barn drying. The advantages of such a system include an improved yield of dry matter (up to 15%) and a greater feeding value.

Barn-drying systems have two features: first is the *ventilation* of the stored bales and second is the use of heated air for *drying*. Good ventilation keeps respiration losses

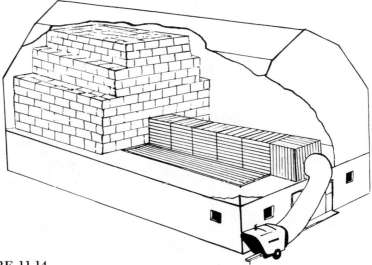

FIGURE 11-14
A fan heater and duct system in the barn allow part of the hay curing to be carried out in storage. This reduces field losses and increases hay quality.

to a minimum. It also prevents movement of water by convection (caused by heating) from one part of the stored crop (usually the center) to the top. If some loss is acceptable, heat generated by respiration may be used to dry the crop by forcing otherwise unheated air through spaced bales. Where the moisture content is about 30%, the loss of dry matter using this method can be about 5%. The type of barn which permits the use of heated air for drying is shown in Figure 11-14. There must be a balance between the atmospheric temperature and the exhausted air. If the air leaving the crop is too hot, it will not be saturated with moisture vapor. Drying is not efficient and the hay at the bottom of the barn may be overdried. On the other hand, if the air emerging is fully saturated and cooled, it will deposit water in the top layers of the hay. Exhaust air during the early drying stages should have a relative humidity of about 90%.

The same principles and equipment (fans and heater units) for drying in the barn can be applied to wagon drying. The entire wagon is covered with a tarpaulin, under which the air is forced. This system is very versatile, but extra wagons are needed to avoid a harvest bottleneck. This method may be used for chopped hay as well as for bales.

Loose Hay Systems

Some of these systems are simply mechanized and unified forms of the long-hay harvesting method described previously. Sweep and stack systems, which use a cage to form the stack, fall into this group. However, many of the stack-wagons which are available chop or break up the hay in some way (Figure 11-15). It is very important that stack-making equipment produce stacks which are uniformly compacted. This is difficult, since even the compression-type stackers are likely to have some poorly packed regions which will settle into depressions that catch and hold water. While it is important that the operator be skilled in the use of stack-wagons, it is usually impossible for him to see the exact shape of the stack he is producing. What type of stack was built becomes evident some weeks later. In all cases, the top of the stack must be well rounded, with no depressions. Poorly topped stacks should be finished by hand, using long hay.

Herbage moisture contents of up to 25%, or in some cases a little higher, are satisfactory for stacking long hay. There is, however, a danger of mold formation and heating if damp weather follows shortly after stacking. Dense stacks made from chopped material (moisture content of about 20% at time of stacking) are more likely to heat than stacks made from long hay. On the other hand, loose stacks are susceptible to rain penetration and may suffer losses for this reason. Dry-matter losses are between 5% and 15% for mechanically built stacks.

Wafered and Pelleted Hay

Hay may be processed to produce cubes or pellets. These have the advantage of being dense and free-flowing, and hence are easy to handle, transport, and store in bulk. They may be moved by using some types of grain-handling equipment. Wafers or cubes

FIGURE 11-15
A forage harvester picking up hay from a windrow.
(Photograph courtesy of Deere & Co.)

are formed from chopped hay. They are "bite-sized," being about 38 mm (1.5 in.) square by 50 to 75 mm (2 to 3 in.) long. Wafer bulk density, ranging from 300 to 450 kg/m^3 (508 to 760 lb/yd^3), is two to three times the density of baled hay. Pellets, which are formed from ground hay, lack the coarse roughage required for cattle. The density is high and may be up to twice that of wafered hay.

Both cubes and pellets may be made by machines operating in the field and by stationary equipment. In both cases, the machines are expensive and the power requirement is high. The mobile field units are limited to use in legume crops in climates which allow quick drying in the windrow to 10% to 12% moisture content. The field cubing of alfalfa is an accepted practice in the dry climates of southwestern, western, and northwestern United States. There the machines are popular, because field losses are lower than with conventional baling equipment, and animal intake of the cubes is high. Also, since cubes are easy to handle, they may be mixed with silage or concentrates to form a complete ration.

Hay Additives

Additives are intended to prevent the development of molds in hay having a higher moisture content than is ideal (30% to 50%). They may also reduce respiration, which would otherwise make use of sugars in the herbage. There are a number of substances which may be used without adversely affecting the feed value of the hay. Propionic acid has been successful in trials in Britain, sulfur dioxide is being tested in Canada, and

TABLE 11-1
Hay-making losses under fair conditions

Cause of loss	Dry matter lost, %
Respiration: Field	Up to 10
Store	5
Weathering in field	10
Mechanical losses in handling	About 10
Fermentation in the stack	0–15
TOTAL	15–35

potassium carbonate is used in Australia. The problem in all cases is that, following their application, there is little movement of these chemicals in the hay. The wider use of such additives will depend on the development of suitable equipment which will give uniform application of small amounts of these chemicals to the hay.

SUMMARY

The object of hay making is to conserve herbage dry matter in such a way that a maximum amount of the plant leaf is retained. A bright green color is taken to indicate that the carotene content is high and that the protein has not been broken down or lost. The time of cutting is at about 1/10 bloom stage for legumes and at anthesis for grasses. Hay should be dried as rapidly as possible and collected from the field when it has a moisture content of about 25% for long, loose, or baled hay, 20% for chopped or shredded hay, and 5% for field cubed or pelleted hay.

Table 11-1 summarizes the types of losses which might be expected in hay making and gives some indication of their magnitude. Table 11-2 shows the importance of handling the crop in such a way that it has a minimum exposure to rain. Even when the hay is well cured and stored, carotene losses due to chemical breakdown will increase in storage. Little carotene will remain after the hay is 1 year old.

TABLE 11-2
Percentage of digestible protein lost due to rain when making hay

Condition	Digestible protein lost, %
Barn dried, no rain	14
Field dried, no rain, some leaf shedding	33
1–2 Showers of rain	28
5–6 Showers of rain	50

(Source: J. Dairy Sci. 22:889–980)

DEHYDRATION

"Dehy," dehydration, artificial drying, or forage drying, is not a new process. It was first used in Europe and Britain in the 1920s. Many thousands of driers were operated during World War II, when concentrated animal feed was rationed or not available. Production costs for this method are high, and it is only where the price of fuel, such as natural gas, is low that dehy products are able to compete with other feeds for farm animals.

Alfalfa is the main forage material dried in North America. In the last 20 years, an increasing amount of dehydrated bermudagrass has been produced in the southeastern United States, while a few plants in Washington dry orchardgrass and perennial ryegrass, sometimes in a mixture with ladino clover.

Worldwide studies of grass-drying methods have led to the development of many dehydration systems. Drying temperatures may be varied widely (from 80° to 800°C or 175° to 1,475°F), provided exposure time is adjusted accordingly (30 min down to 30 sec). High-temperature machines are the most widely used. When using high-temperature machines, the crop is chopped (frequently in the field) and carried through the machine's rotating drum in a hot airstream. The material may be carried once or three times along the visible length of the drum. The weight of the individual pieces of plant material determines how quickly they are carried through the drum. Pieces of leaf, which are light, dry quickly and pass through the drum in as little as 20 sec. The wet, heavy stems move through the drum slowly, taking 2 min or more. From the drum, the material is passed into an expansion chamber to separate the herbage from the humid air. The dried material is passed through a hammer mill, cooled, and compressed into pellets.

The development of low-temperature grass driers has received attention because of the current need for fuel economy. The low-temperature systems call for a longer drying time, resulting in a product with a slightly lower digestibility of crude protein than that resulting from high-temperature drying (see Table 11-3). There is, however, a considerable cost savings, and, where available, lower-grade heat sources can be used.

Artificial drying is the most expensive, but by far the most efficient, way of storing forage. Table 11-4 compares results from feedstuff analyses and digestibility determina-

TABLE 11-3
Digestibility of crude protein in alfalfa samples dried at different temperatures

Drying gas temperature at inlet		Digestibility of crude protein
°C	°F	%
80	75	76
130	266	76
240	464	75
335	635	78
450	842	81
600	1,112	83

TABLE 11-4
Composition and digestibility of fresh and artificially dried alfalfa

	Dry matter, %	
	Fresh	Dehy
Ether extract	2.09	2.79
Crude fiber	21.95	21.92
Crude protein	17.58	17.83
Nitrogen-free extractives	44.86	45.92
Digestibility	74.4	72.3

tions for fresh alfalfa with those from a dehydrated alfalfa sample. While the differences in the values recorded favor the fresh material slightly, they are small and could well fall within the expected range of experimental error. The ascorbic acid (vitamin C) in green herbage is entirely lost in the dehydration process, since it is readily destroyed by heat. Carotene, on the other hand, is almost entirely retained. A 90% recovery may be expected when the carotene contents of fresh alfalfa and dehy material are compared.

In view of the excellent conservation characteristic of the dehy method, every effort should be made to reduce production costs. Attention has been drawn to the use of low-temperature drying systems. Similarly, costs can be reduced by doing some drying in the field. This may be achieved by cutting the crop and leaving it in a windrow for 6 to 8 hr before picking up, chopping, and transporting to the drying factory. Recent work in Holland indicates that heat treatment of the uncut crop is an effective way of reducing moisture content for dehy, hay, or wilted silage production. Flaming the crop at 70°C (158°F) was not successful, since the leaves were damaged, while the lower stems remained undried. Steam treatment, in which water is dispersed directly into a flame to give a mixture of steam and hot gas, provides uniform and efficient heat transfer when applied in a jet to a standing crop. It is estimated that the use of steam drying in the field could reduce operating costs for the dehy process by 50%. The capital cost per tonne of dried crop is also significantly reduced, since the processing time is decreased.

FURTHER READING

Fulkerson, R. S., D. N. Mowatt, W. E. Tossell, and J. E. Winch, 1967, Yield of dry matter in vitro-digestible dry matter and crude protein of forage, Can. J. Plant Sci. 47:683–690.

Kilcher, M. R., and D. H. Heinrichs, 1961, A grass alfalfa mixture compared with cereal grains for fodder under semi-acid conditions, Can. J. Plant Sci. 41:799–804.

Polan, C. E., T. M. Starling, J. T. Huber, C. N. Miller, and R. A. Sandy, 1968, J. Dairy Sci. 51:1801–1805.

Sheard, R. W., and J. E. Winch, 1966, J. Brit. Grassland Soc. **21**:231–237.

Skidmore, C. L., ed., 1973, Proceedings of the 1st International Green Crop Drying Congress, Oxford, England, p. 536.

Smith, A. D., A. Johnston, L. E. Lutwich, and S. Smoliak, 1967, Can. J. Plant Sci. 48:125–132.

Watson, S. J., and M. J. Nash, 1960, The Conservation of Grass and Forage Crops, 2nd ed., Oliver & Boyd, Edinburgh.

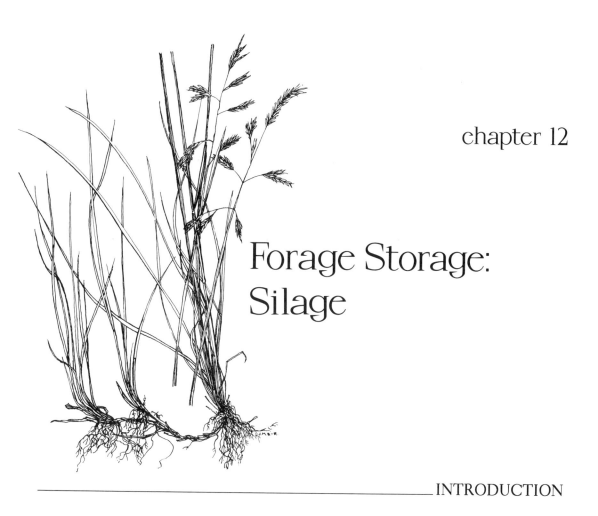

chapter 12

Forage Storage: Silage

INTRODUCTION

The process of ensilage is quite different from the dry storage methods discussed in the previous chapter. When silage is made, the *wet* crop is preserved by *chemical* action. The problem in storing cut green herbage is that, as the material decomposes, it heats up, thus increasing the rate of oxidation and decomposition. The air which is mixed with the forage becomes warm and rises, drawing fresh cold air into the lower part of the heap of green material. This provides a new supply of oxygen, which further speeds the process of decomposition and heating. To prevent decomposition, it is essential to stop air movement. This is done in two ways. First, the green material is compacted so that, initially, it will contain a minimum amount of air and, as a result, subsequent air movement in the herbage will be reduced. Second, the green material is covered or enclosed in a container (called a *silo*). Then two things happen. First, the respiration of the living, green material with the aerobic bacteria uses up all the oxygen in the container, producing carbon dioxide. There will be some increase in temperature at this stage, but if the herbage is compact and the oxygen supply is limited, the process will last only a few hours and the temperature will not rise above 30°C (86°F). At that time, the ensiled material will settle, forcing out the air and gasses it contains. This completes the first stage of silage making.

In the second stage, the anaerobic conditions produced in stage 1 kill the plant cells. As the cell membranes degenerate, exudates containing sugars move out of the plant and become available to the bacteria which are normally present on the plant's surface. The most important of these are the *lactobacilli*, which are able to ferment sugars, first producing pyruvic and then lactic acid. These organisms are anaerobic and function well under acid conditions (down to pH 3.0). They continue the fermentation process until the concentration of lactic acid is 8% to 9%, at which time the pH has been reduced to 4.2 or lower. Under these circumstances, the fungi, bacteria, and enzymes which normally decompose green forage material are inactive, because the conditions are anaerobic (as in the case of the molds) or because the pH is low (many bacteria are inactive at pH 4.3 or lower). It takes 2 or 3 weeks to complete the two stages of ensilage. The production of anaerobic conditions will take only a few hours if the material is compacted and carefully enclosed. In the subsequent 3 or 4 days, the lactic acid-forming bacteria increase in numbers from the few which initially are present on the surface of growing plants everywhere in the world, to the hundred million or so per gram of forage which are needed to make silage. The remainder of the time is spent in the production of lactic acid and other weaker acids, such as propionic, acetic, formic, and succinic. As the pH declines to 4.3, the bacteria which produce acids other than lactic acid become inactive and the proportion of the latter increases rapidly. Once the pH is 4.2 or lower, preservation is permanent and no change will take place in the conserved material, provided the pH does not increase again. While most of the silage made on the farm is used in the same year, reports are not uncommon of individuals satisfactorily using 2- or 3-year-old material.

PROCESS OF BACTERIAL FERMENTATION

The fermentation process which leads to the conversion of plant sugars into pyruvic and other acids is carried out by many bacteria and differs in its chemical characteristics from alcoholic and homolactic fermentations. These differences depend entirely on the metabolism of pyruvic acid. The anaerobic bacteria concerned include *Escherichia, Salmonella, Shigella, Proteus, Yersinia,* and *Vibrio.*

A generalized diagram showing the end products is given in Figure 12-1. These reactions are collectively known as *mixed-acid* fermentation. While the fermentation of acetic acid, ethanol, and formate (or CO_2 and H_2) from glucose is a balanced reaction and could, theoretically, yield these substances as the only end products, this rarely happens. Many of the organisms which are involved in mixed-acid fermentation are able to reduce pyruvic acid to lactic acid. A substantial amount of lactic acid is always produced, the exact amount depending on the environmental circumstances.

As well as mixed-acid fermentation, *butanediol* fermentation is carried out by bacteria which are members of the genera *Aerobacter, Serratia, Erwinia,* and some species of *Bacillus* and *Aeromonas.* Here, 2,3-butanediol is formed along with the typical products of mixed-acid fermentation. Bacteria of the genus *Clostridium* will ferment acetic and butyric acids. Here, also, the end products are numerous and the ratios between them vary widely. Many of the *Clostridium* bacteria can also ferment

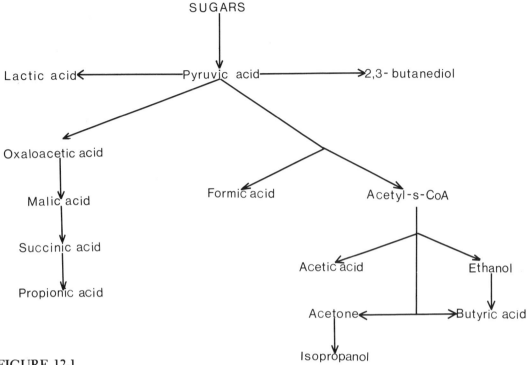

FIGURE 12-1
Bacterial fermentation of plant sugars to form pyruvic acid and a mixture of other acids.

amino acids. This brief outline of the mixed-acid fermentation process indicates the biochemical complexity of the system and the extreme physiological specialization of the organisms involved.

SILAGE-MAKING PROBLEMS

What can go wrong in the basic silage-making process? For what reasons may we encounter problems? There are three fundamental causes for the production of bad silage:

1. Too much air in the green material.
2. Too much water in the herbage.
3. Too little carbohydrate available for fermentation.

There are, however, no *absolute* levels of air, water, and carbohydrate for ensiling, but the proportions of these three substances in the herbage determine the nature of the silage produced. For example, if the water content is high and the proportion of plant sugars is low, the silage produced will not be fully preserved. However, a reduction in the water content can result in the production of excellent silage even though the plant sugars remain low. Let us now examine, in turn, each of the three aforementioned fundamental characteristics of material to be ensiled and see how they influence the end product.

Too Much Air

The amount of air present in the material placed in the silo is important to the first, or respiration, stage of silage making. If the herbage is too "open," or if the container permits the entry of too much air, the respiration stage is prolonged. This will break down carbohydrate material which might otherwise be available for fermentation into lactic acid or for animal nutrition. Also, the temperature of the herbage will rise and may become high enough to char the silage, giving it a burnt or "toffee" smell. Under these circumstances, proteins and carbohydrate substances are denatured, and the silage so produced is unpalatable to livestock.

There also are problems associated with the presence of air during the second stage of silage production. If anaerobic conditions do not exist when fermentation starts, butyric acid-producing bacteria, which require oxygen, become active. In contrast with the lactic acid-producing bacteria, these organisms produce a weak acid with the unpleasant smell and taste of rancid butter. Even when silage has been successfully preserved and the pH reduced to 4.2 or lower, air moving into the material may result in lactic acid being converted to butyric acid, as shown in the following formula:

$$2(C_3H_6O_3) \rightarrow C_4H_8O_2 + 2H_2 + 2CO_2$$
$$\text{lactic acid} \qquad \text{butyric acid}$$

The weaker acid will fail to reduce the pH, thus allowing molds and fungi to become active. They attack the protein molecules and produce ammonium ions, which increase the pH even more.

Too Much Water

If the crop has a high moisture content when it is ensiled, there is danger that the second stage (i.e., the fermentation) will be prolonged or that carbohydrates will be leached from the silage. The additional water also dilutes the lactic acid and delays the time when the pH will fall to 4.2, and some material, which might otherwise have been available for animal nutrition, will be used to produce lactic acid. Such a loss may be avoided if the herbage is allowed to wilt in the field. As the water is removed the osmotic pressure of the fluid part of the cell content of the forage increases. This has the advantage of inhibiting the breakdown of protein by proteolytic organisms. The conversion of protein into nonprotein nitrogen compounds ceases when the dry matter content is 35% or at a pH value below 4.2.

TABLE 12-1
Percentage of sugar content on a fresh-weight basis for green forage material

	Sugar by weight, %
Alfalfa	1.1
Clover	0.75
Oats in head	2.1
Corn	5.0

Too Little Carbohydrate

This condition is frequently associated with the high water content of direct-cut forage. Also, many legumes do not contain enough carbohydrate material to produce all the lactic acid required for preservation. Furthermore, the carbohydrates available in the herbage may be low because the respiration stage was prolonged.

While the exact amount of lactic acid needed can only be determined when the moisture content of the herbage is known, at least 2% by weight of the green material should consist of plant sugars. As is shown in Table 12-1, the sugar content of oats in head is satisfactory, while the value for corn is more than adequate. Alfalfa and clover both give values below the minimum sugar content required. The percentage of sugar present, on a green matter basis, may be increased by wilting the herbage in the field or by adding molasses or other substances in the silage.

METHODS OF MAKING SILAGE

When making silage, we place three different substances in the silo: (1) plant material, (2) air, and (3) bacteria. As we have seen, the composition of the plant material, the initial amount of air present (or that which subsequently enters), and the type of bacteria present jointly determine the kind of silage produced. The farmer can readily control the first two of these three factors. The way in which this is done determines the nature of the silage produced and may result in silage of three distinct types: (1) direct-cut silage, (2) wilted silage, and (3) low-moisture silage.

Direct-Cut Silage

As the name implies, direct-cut silage is made by cutting and immediately ensiling the crop. It has three advantages over alternative methods. First, it may be made under any weather conditions, even when it is raining. Second, fewer machines are required than for other silage-making methods (Figure 12-2). Third, the crop is cut at the time when there is an optimum balance between yield and quality (late bud or early bloom stage for legumes and early heading stage for cereals and grasses).

FIGURE 12-2
A self-propelled forage harvester will finely chop the entire crop and make it ready for the silo. The moisture content of the herbage will be determined by the stage of maturity of the crop. (Photograph courtesy of Sperry-New Holland)

One disadvantage of the direct-cut method is that the moisture content of the herbage material is high. It may be over 70% by weight and is frequently as high as 80%. With this level of moisture, the silage may be sour and dark olive in color. Frequently, there is also a nutrient loss due to the seepage of liquids from the bottom of the silo. Where winters are cold, freezing may be excessive. It is possible to overcome these difficulties by direct-cutting mature herbage which has a lower moisture content (about 65%). At that stage, the forage yield will be higher and the quality lower than at the optimum harvest date.

An alternative method of overcoming the silage-making problems encountered with direct-cut silage is to add material that either acts as a preservative or absorbs moisture. Examples of such substances are molasses, which adds fermentable carbohydrates, or corn and cob meal, which adds carbohydrates and absorbs moisture. The addition of 30 to 50 kg (66 to 110 lb) of molasses per tonne of silage will substantially improve lactic acid production so that the silage is sweet and of a normal green color. About 75% of the feed value of the molasses will remain for animal nutrition. The addition of 75 to 100 kg (165 to 220 lb) of corn and cob meal per tonne of ensiled material will also improve silage quality. In this case, about 80% of the feed value of the meal is retained. The addition of additives involves an expense and some management problems. This, combined with difficulties encountered due to freezing, has resulted in limited use of this method.

FIGURE 12-3
This self-propelled forage harvester will pick up green material from the windrow and chop it ready for storage in the silo. (Photograph courtesy of Sperry-New Holland)

Wilted Silage

As this name suggests, wilted silage is made by allowing the forage to dry in a windrow for 2 or 3 hr. This will reduce the moisture content to between 60% and 70%. More equipment is required than for direct-cut silage (Figure 12-3). Plant material with a moisture content of 65% or less is light and therefore requires chopping and compacting to ensure that the amount of air in the material is not excessive. For wilted silage, the material is usually cut to a length of 13 mm (0.5 in.). The length of cut is far more important than for direct-cut silage, where the heavier material, with a higher water content, compacts readily.

This type of silage has the advantage of being free from the problems of odor, freezing, and seepage. Also, fermentation is less extensive than for the direct-cut method, so the amount of carbohydrate material used for lactic acid production is less. A pH of 4.5 is satisfactory for conservation. While preservatives are not required, cereal grain may be added to wilted silage to increase the energy content.

Low-Moisture Silage (Haylage)

Low-moisture silage is produced in much the same way as wilted silage. There are, however, two important differences. First, the herbage is retained in the windrow until the moisture content falls to about 50%. Second, material of this type must then be cut to a length of 7 mm (0.25 in.). Compaction is *very* important, since, with a moisture content between 40% and 60%, bacterial activity, and consequently mixed-acid fermentation, will be limited. There will be little acid production and the pH may not fall to any great extent. Hence, it is essential that there be as little air as possible present in the herbage initially and that its subsequent entry be prevented.

Low-moisture silage should be stored in a gas-tight or tower silo. If a horizontal silo is used, it should be sealed well, using high-moisture silage to cover the top. The low-moisture silage method has the advantage of incurring lower dry-matter losses than any other silage-making or hay-making method.

CHEMICAL ADDITIVES

As well as the plant-product additives (such as molasses and cereals) which were mentioned previously, a number of chemicals, usually acids, can be used to improve silage production methods. The most widely used method of this type is the *AIV acid* system. Here, a mixture of sulfuric and hydrochloric acids is added to the crop as it is put into the silo. The advantage of this method is that the losses in the first, or respiration, stage of silage making are low. A low pH does not inhibit the activities of the lactic acid bacteria, but does suppress other bacteria and molds. One disadvantage of artificial acidification is that the AIV acids are corrosive and unpleasant to handle, damaging both equipment and clothing. Also, the animal intake of the silage so produced is often low, and the livestock may suffer digestive upsets.

This acidification method is usually used with high-moisture silage. A liter of concentrated acid is mixed with 11 liters (l) of water. Six liters of this mixture is then applied to 100 kg of grass. About 8 l of the mixture is used for 100 kg of legume material. Formic acid, which has been used in a similar way, has been shown to be particularly effective in preventing protein breakdown. Table 12-2 shows data from a silage making and feeding experiment in which formic acid-treated silage is compared with untreated silage. Both the fermentation temperature and the amount of nitrogen in a nonammonia form are reduced, and animal-weight gains are increased.

The most serious objections to artificially acidified silage are the decrease in animal

TABLE 12-2
Formic acid-treated silage fed to 24 steers weighing 180 to 270 kg (396 to 594 lb)

	Max temperature of herbage °C	*Dry matter %*	*Protein % of dry matter*	*NH_3 as a % of dry matter*	*Silage intake kg per animal for 69 days*	*Liveweight gain kg per animal per day*
Formic acid treated	49	33	14.6	10.2	1,390 (3,058 lb)	0.75 (1.50 lb)
Untreated	53	33	14.1	13.4	1,360 (3,058 lb)	0.50 (1.0 lb)

(Source: Alberta Agriculture publications)

intake and the digestive problems that accompany its use. This has led to the examination of other substances which will preserve the crop at a high pH. The greatest success has been obtained using formaldehyde. This chemical is a sterilizing agent which will kill bacteria and molds and, at the same time, combine with protein to prevent protein decomposition. However, if air is subsequently allowed to enter the sterilized herbage, the herbage will become contaminated with bacteria and molds, and decomposition will start. Formaldehyde is a useful adjunct to the production of low-moisture silage, for then a well-sealed silo is essential in any case. A mixture of formaldehyde and formic acid has the advantage of giving some protection against the secondary entry of bacteria and molds. The satisfactory use of all additives usually depends on the availability of suitable application methods.

HARVEST EQUIPMENT

Mowers, rakes, and conditioners are used to prepare windrows in the same way as previously described for hay (Figure 12-4). Forage harvesters are then used to chop and load the herbage, either from the windrow or as a direct-cut crop. The flail and double-chop types of harvesters produce a variable cut that is not satisfactory for silage. Radial knife and cylinder harvesters (the cylindrical cutter head being the most popular) are more expensive, but give a cut of a uniform length.

FIGURE 12-4
Hay-making equipment can also be used to prepare windrow for wilted or low-moisture silage.
(Photograph courtesy of Sperry-New Holland)

FIGURE 12-5
A large-capacity forage wagon unloading silage into a horizontal silo. (Photograph courtesy of Sperry-New Holland)

FIGURE 12-6
A self-unloading forage wagon working with a blower to elevate silage to the top of a vertical silo. (Photograph courtesy of Sperry-New Holland)

The chopped herbage is blown into a self-unloading wagon. The unloading process depends on the type of silo used. For the horizontal or trench silo, the material may be dumped by tipping or moved through the endgate by a floor conveyor (Figure 12-5). For a vertical silo, the same methods or self-unloading through a wagon chute may be used to move the herbage to a blower that will elevate it to the top of the silo (Figure 12-6). The main variations in silage-making systems are in hauling and unloading methods. The speed of a silage-making operation (an important factor in determining cost) is dependent on many features. The most important are the power available to drive the harvester and the length to which the forage is cut.

TYPES OF SILOS

The many types of silos used may be grouped into two categories. Either they are of the upright, tower, or vertical type, or they are of the horizontal, trench, or bunker type. Both types are constructed from a wide range of materials, including steel, wood staves, and concrete. Handling silage in horizontal silos may be more difficult and expensive than in the vertical type. However, most farmers favor horizontal silos because the original capital cost is substantially lower than that of a vertical silo. Horizontal silos are frequently unloaded by using a front-end loader. Special cutter reel unloaders, which cut down the face of the silage and carry the loosened material to a wagon, are also available. Movable gates may also be used to facilitate self-feeding.

Vertical silos can be automatically unloaded by top or bottom unloaders. The top unloader is more popular, being less costly and more reliable than the bottom unloader. However, the bottom unloader is of considerable advantage in a fully sealed silage system. Table 12-3 gives an indication of the ranges of herbage moisture content for different types of silos.

A different type of silo has attracted attention in recent years. This consists of a ground sheet on which the herbage is placed and then covered by a top sheet of polyethylene (500 gauge or thicker). The edges of the two sheets are sealed by using concentric tubing, and a vacuum pump removes the air. In this "vacuum" silage,

TABLE 12-3
Relationship between silo type and the moisture content of the herbage

Type of silo	Forage moisture %
Gas-tight	35–40
Upright	60–70
Trench or bunker	65–75
Stack	
Over 3.5 m (4 ft)	68–73
Under 3.5 m (4 ft)	72–77

(Source: From various publications by Alberta Agriculture and Agriculture Canada)

TABLE 12-4
Dry matter losses in making silage

Cause of loss	Dry matter lost, %
Respiration	3–15
Fermentation	8–30
Mechanical losses	1–4
Seepage	0–3
Top spoilage	4–15
TOTAL	16–67

(Source: From various publications by Alberta Agriculture and Agriculture Canada)

oxidation is reduced to a minimum because of lack of air and the material is also compressed by the polyethylene envelope in which it is contained.

SUMMARY

The types of losses that may be encountered in silage making are listed in Table 12-4. In addition to the respiration and fermentation losses that have been discussed, there are some losses which occur when bringing herbage to the silo and in moving the silage to the animals. Seepage losses occur where high-moisture silage is made, and there is always "top" spoilage found at the top and sides, especially in horizontal silos. The total losses will range from 15% to nearly 70% of the dry matter in the original material.

The extent of the dry-matter loss is an important feature of all storage processes. With the exception of dehydration (dehy), the methods of hay and silage making set out in this and the previous chapter are part of a continuum, rather than being isolated and separate methods. The association between the various storage methods depends on the moisture content of the plant material at the time when it is harvested. Field-dried hay may contain as little as 10% moisture. Direct-cut silage may have a moisture content as high as 80%. Between these two extremes are the moisture contents of all the types of silage and hay discussed. It is possible to relate the moisture contents to dry-matter losses in both the field and in storage. Field losses are high for hay making and low for silage. The reverse is true for losses in the store. The total losses are low when forages are harvested with intermediate moisture contents (40% to 60%), as is shown in Figure 12-7.

Dry-matter losses are always paralleled by a loss of nutrient material. It is necessary to point out the obvious by saying that no stored product is ever any better than the original material from which it is produced. Frequently, it is considerably poorer. Furthermore, since the moisture content of silage is high, it is possible to use silage to fill and content livestock, and yet provide them with less nutrition than when feeding hay. Table 12-5 shows that the average daily weight gain for steers at the University of Alberta ranch was highest when they were fed hay and indicates the importance of

FORAGE LOSSES

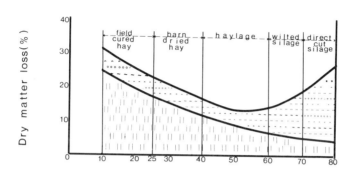

FIGURE 12-7
Both field and storage losses must be considered when evaluating forage storage systems. (Source: Canadian Forage Crop Symposium, 1969)

dry-matter intake. On the other hand, the nutrient losses from field curing hay are much higher than those recorded for silage making. Data from Vermont (Table 12-6) illustrate this, showing losses two to three times as great for hay as those for silage. This difference will be most marked where frequent rains or high humidity are likely. Both of these conditions make the production of good-quality hay difficult.

TABLE 12-5
Dry matter intake and daily weight gains for steers fed three kinds of conserved material

	Dry matter content %	Dry matter intake per unit body weight		Average daily gain	
		kg per 100 kg	lb per 100 lb	kg	lb
Unwilted silage	18	1.6	1.6	0.05	0.1
Wilted silage	25	2.0	2.0	0.5	1.1
Hay (alfalfa–bromegrass)	82	2.3	2.3	0.7	1.5

TABLE 12-6
Percentage of nutrient losses occurring in alfalfa and timothy conserved as hay and silage

Feed constituent	Field-dried hay	Molasses silage
Dry matter	15.8	8.2
Ash	21.5	6.8
Crude protein	22.8	8.0
Crude fiber	11.8	0.6
Nitrogen-free extract	13.8	13.9
Ether extract	28.4	15.6

(Source: Vt. Agr. Expt. Sta. Bul. 434)

FURTHER READING

Barnett, A. J. G., 1954, Silage Fermentation, Academic Press, New York.

Essig, H. W., 1968, Urea limestone-treated silage for beef cattle, J. Anim. Sci. **27**: 730.

Lancaster, R. J., 1966, Silage: The research view-point, Proc. 20th Conf. N.Z. Grass Assoc., pp. 154–62.

Langston, C. W., H. Irwin., C. H. Gorden, C. Bouma et al., 1958, Microbiology and chemistry of grass silage, U.S.D.A. Tech. Bull. 1187.

Miller, R. W., L. A. Moore, D. R. Waldo, and T. R. Wrenn, 1967, Utilization of corn silage, carotene by dairy calves, J. Anim. Sci. **26**:624.

Roffler, R. E., R. P. Niedermeier, and B. E. Baumgardt, 1967, Evaluation of alfalfa-brome forages stored as wilted silage, low moisture silage and hay, J. Dairy Sci. **50**: 1805–13.

Whittenbury, R., P. McDonald, and D. G. Bryan-Jones, 1973, A short review of some biochemical and microbial aspects of ensilage, J. Sci. Food Agr. **18**:441–46.

Woodward, T. E., 1944, Making silage by the wilting method, U.S.D.A. Leaflet 238.

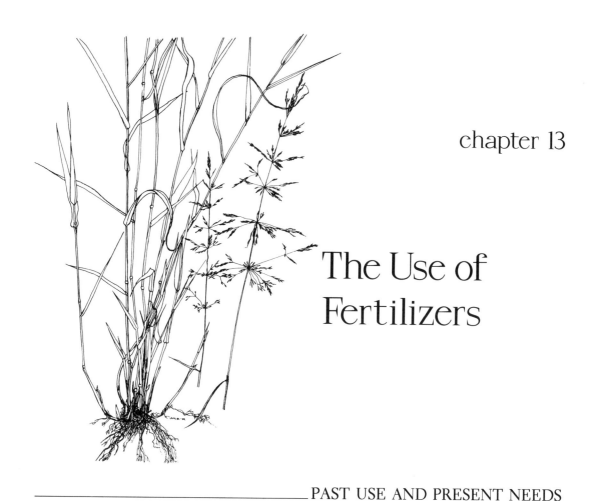

chapter 13

The Use of Fertilizers

PAST USE AND PRESENT NEEDS

The present world energy situation places forage crops in a position of vital importance for the maintenance of livestock and consequently for our own welfare. Despite this, farmers seldom devote adequate time and attention to the conservation, cultivation, and fertilization of their fodder crops. These are frequently relegated to poor or marginal land. There are two reasons for this. First, forage marketing systems, where they exist at all, are far from satisfactory. Value is often determined entirely by quantity, with no premium being paid for quality. Consequently, the more readily marketable cash crops are given priority in farm production planning. Second, where the herbage is intended for use on the farm, a low yield from a large area is often more acceptable to the farmer than the expense of a fertilizer application.

Increases in the capital cost of land in recent years make the economics of such thinking quite unsatisfactory. Profit per hectare (acre) must be maximized by combining the highest possible yield of good-quality forage with efficient and full utilization. Today, the rates of fertilization used for forage crops in the North American continent are grossly inadequate. This, combined with the location of pasture and hayland on poor soils, means that the possibilities for large increases by improving soil fertility and pH levels are enormous.

Turning to the tropical areas of the world, we find that, relative to needs, fertilizer applications to forage crops are almost nonexistent. In this case, phenomenal growth increases would be possible if a "package" of forage practices (including fertilization) could be adopted as an entity. The approach used in India with cereal grain production should be applied to forage crops. There is no doubt that the development of satisfactory fertilizer practices here at home or in Third World countries could increase world food supplies more than any other improvement in agricultural practice.

The effect of fertilizers on grasslands was first studied in Europe in the last years of the nineteenth century (see Chapter 2). The use of basic slag and ammonium sulfate indicated the importance of nitrogen and phosphorus; experiments started in 1916 at Hohenheim added potassium to the list of essential major nutrients. Fertilizers were not used on grasslands in North America before 1930. Soil fertility studies conducted between 1920 and 1930 show the importance of lime and phosphates for legume growth. Also, a great many trials conducted since then show that, as long as soil moisture was adequate, nitrogen increased the yield of grasses in haylands and pastures. It was, however, not until high-yielding grass–legume mixtures such as orchardgrass and ladino clover, and smooth bromegrass and alfalfa, and bermudagrass and clover were grown in the 1940s that the full value of fertilizers for forage production was appreciated. Then, as soil-testing technology developed, forage fertilization became an accurate science.

NUTRIENT REQUIREMENTS

In general, forage crops remove from the soil considerably more nutrients than do cereal grains. With forage crops, almost all the dry matter the plant produces is removed. With cereal crops, the straw is frequently returned to the soil, thereby helping to maintain soil fertility. A brome–alfalfa hay crop yielding about 4 T/ha (8 tons/acre) of dry herbage will remove from the soil 80 kg (176 lb) of nitrogen, 8 kg (17.5 lb) of phosphorus, 34 kg (75 lb) of potassium, and 3.5 kg (7.75 lb) of sulfur. Obviously, if this process continues year after year, the voracious appetite of the forage species will make even the most fertile soil deficient in these major nutrients. For some nutrients, this condition already exists on many permanent pastures. Nitrogen deficiencies are almost universal. Although nitrogen mineralization occurs at different rates in different soils, there is no soil type which can, under grass, supply adequate amounts of available nitrogen for sustained high yields.

Grasses

Nitrogen. A grass takes up nitrogen as a nitrate through most of its life. Only the young seedling preferentially absorbs nitrogen in the form of ammonia. When a nitrate is absorbed, it is converted into ammonia in the plant, since this is the first step in the production of plant protein. All forage grasses may be regarded as gross nitrogen "feeders," rapidly taking up nitrogen from the soil. To assure good grass growth, abundant supplies of nitrogen must be available *throughout* the growing season. Conse-

TABLE 13-1
Effect of ammonium nitrate fertilizer on first-cut yields of bromegrass hay at four locations

METRIC

N application (kg/ha)	Dry-matter yield (kg/ha)			
	Scott	Elstow	Whitewood	Loon River
22	1,672	1,974	1,960	1,592
45	1,998	2,639	2,664	2,229
90	2,493	3,241	3,318	2,901
180	2,856	3,640	3,732	—
Nil	1,023	1,463	1,177	854

ENGLISH

N application (lb/acre)	Dry-matter yield (lb/acre)			
	Scott	Elstow	Whitewood	Loon River
20	1,488	1,771	1,744	1,417
40	1,778	2,349	2,371	1,984
80	2,194	2,884	2,953	2,581
160	2,542	3,240	3,321	—
Nil	910	1,302	1,048	760

(Source: Canadian Forage Crops Symposium, 1969)

quently, the nitrogen fertilization of cool-season grasses will give greater changes in yield and forage composition than the addition of any other major nutrient element. This statement is supported by a voluminous literature. The yields of the first hay cut recorded in one such paper are given in Table 13-1. The second cut from these trials, in the same season, also shows substantial yield increases where fertilizer was applied.

A number of authors have summarized the responses of a range of warm-season grasses to nitrogen applications. Pensacola bahiagrass gives a linear dry-matter increase in response to increasing dressings up to 575 kg of nitrogen/ha (512 lb/acre) in Alabama and up to 450 kg/ha (400 lb/acre) in Florida. In Georgia, yield increase rates started to slow with applications higher than 200 kg of nitrogen/ha (178 lb/acre). Dallisgrass, grown in Queensland, Australia, showed a similar response pattern. Data for Coastal bermudagrass, grown under irrigation in California, show that optimum yields of dry matter are obtained in response to applications of between 950 and 1,400 kg of nitrogen/ha (846 and 1,246 lb/acre). Nitrogen applications higher or lower than these levels give lower yields.

High nitrogen applications alone will not give favorable yields if other growing conditions are unsatisfactory. This is equally true for cool- and warm-season grasses. As well as adequate soil moisture, high carbohydrate reserve levels during the fall and winter are essential for maximum responses to nitrogen fertilization in the following year. Nitrogen applications result in the use of photosynthates for the production of top growth, and under these circumstances, stored carbohydrates are also used for growth. Such withdrawals and use have a secondary effect which is important and advantageous. The utilization of carbohydrate material for the production of top growth precludes the development of roots and rhizomes for storage purposes. Thus, nitrogen fertilizers can reduce the *sod-bound* condition which is frequently associated with old unfertilized stands of grass like bromegrass.

As well as increasing dry-matter production, nitrogenous fertilizer applications will increase the protein content in both cool- and warm-season grasses. Such increases do not occur uniformly throughout the life of the plant, but are greatest in the juvenile stage and decline rapidly as the plant reaches the hay stages (Figure 13-1). Consequently, where feed with high levels of protein is important, the pasture should be grazed in the juvenile stage, using rotational grazing. However, even if fertilized plants are cut at the hay stage, the total yield of protein per hectare is higher than that from an unfertilized area, since the total dry-matter production is greater. A study of nitrate nitrogen responses in smooth bromegrass in North Dakota over a 22-year period showed that rates of 66 kg of nitrogen/ha (60 lb/acre) increase the yield of dry matter by 214% and that rates of 133 kg/ha (120 lb/acre), which gave a yield increase of 257%, are economical if the additional crude protein in the herbage is utilized in such a way as to replace an expensive protein supplement in the diet.

There is evidence to indicate that, for some grass species, a split application of nitrogen fertilizer (equal parts in the spring and after the first harvest) results in a

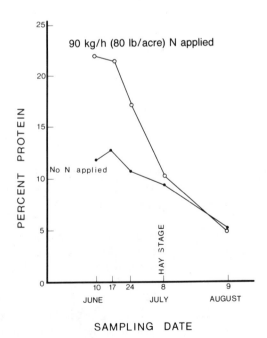

FIGURE 13-1
Protein content of fertilized and unfertilized bromegrass. (Source: Canadian Forage Crop Symposium, 1969)

TABLE 13-2
Percentage of nitrogen recovery from three grasses as influenced by rate and frequency of nitrogen applied

Split annual fertilizer (N) treatment				recovery of N, %					
				Reed canarygrass		Creeping foxtail		Bromegrass	
kg/ha		lb/acre		Cutting		Cutting		Cutting	
				1st	2nd	1st	2nd	1st	2nd
56 +	56	50 +	50	13	9	8	8	14	22
112 +	112	100 +	100	17	11	14	10	17	20
224 +	224	200 +	200	19	15	14	12	25	27
448 +	448	400 +	400	17	8	11	9	13	17

(Source: C. L. Hanson, J. F. Power, and C. J. Erickson, 1978, Agronomy J. 70. By permission of The American Society of Agronomy, Inc.)

greater total recovery of nitrogen than does a single application of the same total amount at the beginning of the season. Table 13-2 shows that the recovery of nitrogen from the second harvest is lower than that from the first harvest for reed canarygrass and creeping foxtail, but is consistently higher from the second harvest for smooth bromegrass.

Apart from the economic losses due to leaching that might accompany high nitrogen applications to a pasture, such dressings may also cause digestive disorders in grazing animals. The most common of these is nitrate poisoning. This occurs in cattle when their diets contain an excess of 0.35% to 0.45% nitrate nitrogen. This condition is, however, influenced by other components of the animal's diet (e.g., the proportion of readily available carbohydrate), and levels of nitrate nitrogen as low as 0.14% have been regarded as potentially dangerous. Applications of a phosphate together with the nitrogenous fertilizer usually result in nitrate levels in the herbage lower than those when nitrogen alone is used, so the use of phosphate may offset nitrate poisoning. The problem of nitrate poisoning is also discussed in Chapter 10.

Phosphorus. The percentage of phosphorus in grasses ranges from 0.14% to 0.50%. While there is no doubt that soil fertility influences the phosphorus content of herbage tissue, phosphorus supplementation of animal feed is best accomplished by adding minerals to livestock rations, rather than by soil fertilization. There is, however, some evidence that the effect of an accumulative decline in P_2O_5 is exhibited by several soils. Figure 13-2 shows a marked increase in yield following the third annual application of 90 kg of P_2O_5/ha (80 lb/acre), together with the same amount of nitrogen.

Potassium. Except where marked soil deficiencies exist, both warm- and cool-season grasses are considerably less responsive to potassium than to nitrogen fertilization. However, in the southeastern United States, a fertilizer application of the ratio

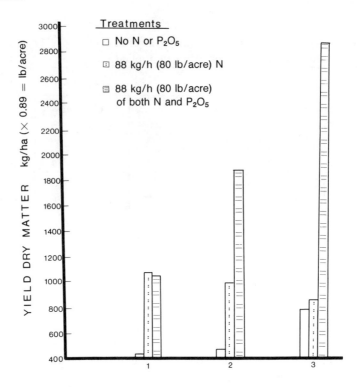

FIGURE 13-2
Effect of annual applications of phosphate on an old bromegrass stand. (Source: Canadian Forage Crop Symposium, 1969)

of 4–1–2 ($N-P_2O_5-K_2O$) is generally recommended for Coastal bermudagrass to avoid P–K stress. Intensive long-term use of nitrogen fertilization may result in the phosphorus and potassium interaction being important.

There is evidence of differential potassium requirements among grasses. For example, orchardgrass has a higher potassium requirement than Kentucky bluegrass. Such considerations are important in relation to the use and evaluation of forage mixtures.

Sulfur. Sulfur deficiencies may occur in all grasses grown on sandy soils with a low organic matter content. However, because most grass species have a low tissue requirement for sulfur and are not particularly sensitive to low sulfur levels, problems are not common. Coastal bermudagrass, which does not normally require sulfur, will give increased yields following sulfur applications when grown on the Eustis Sands of the Carolina sandhills. Obviously, soil conditions influence the plant's needs for sulfur applications.

TABLE 13-3
Effect of applied nitrogen on nitrogen fixation in established ladino clover

| Nitrogen applied | | Nitrogen in the plant |
kg/ha	lb/acre	fixed from the atmosphere, %
28	25	91
56	50	87
112	100	75
224	200	43

(Source: North Carolina Agric. Exp. Sta. Bull. 383)

Legumes

Nitrogen. In many areas improved establishment and increased yield of alfalfa in the first year of growth result from the use of low rates of nitrogen fertilizer. With this exception, it is only on acid soils, where nodulation and nitrogen fixation are poor, that legumes show any response to nitrogen fertilizers. Neither legume productivity nor forage quality is enhanced by nitrogen applications. There is evidence to indicate that adding nitrogen fertilizer to established legume stands will make nitrogen-fixing bacteria in the root nodules inactive. Table 13-3 illustrates this point, showing that increased rates of nitrogenous fertilizer applied to ladino clover depress the percentage of nitrogen fixed from the atmosphere. Inoculation of legumes with the proper strain of rhizobia (see Chapter 6) is the most effective and economical way to ensure adequate nitrogen nutrition.

Phosphorus. Phosphorus may become unavailable to the plant, especially in soils which are at all acid or of a fine texture. Also, there is little movement of applied phosphorus through the soil profile. These two features suggest that *banded placement* at the time of establishment should be beneficial. A large number of research workers have shown that this is indeed the case. On established stands, the problem is rather more difficult, since most legumes are extremely deep rooted. Soils on which alfalfa has grown for some years frequently have depleted phosphorus levels at depths of as much as 75 to 90 cm (2.25 to 2.75 ft) (Table 13-4). No doubt, the lack of movement of phosphorus down the soil profile, combined with the varied rooting depth of legumes, accounts for the variable results, reported from all parts of North America, as a consequence of phosphate fertilizer applications.

Potassium. Numerous experiments show the beneficial effects of this substance on legume yield, quality, and stand longevity. This is illustrated by the data given in Table 13-5, which show that an application of 187 kg of potassium/ha (166 lb/acre) more than doubled alfalfa yields, averaged over a 9-year period.

Potassium plays an important role in the speed of regrowth for most legumes. Rapid regrowth after harvest makes a major contribution to seasonal production, since

TABLE 13-4
Effect of alfalfa on available phosphorus at varying soil depths

		SOIL DEPTH					
	cm	0–15	15–30	30–45	45–60	60–75	75–90
	ft.	0–0.45	0.45–0.9	0.95–1.35	1.35–1.8	1.8–2.25	2.25–2.7
				Soil P (ppm)			
No alfalfa		27.0	24.0	12.7	9.7	12.6	12.9
Alfalfa		15.5	9.4	5.9	6.1	9.6	10.9

(Source: R. C. Lipps, and R. L. Fox, 1956, Soil Sci. Soc. Amer. Proc. 20: 20–32)

rapid and high light interception is important for maximum photosynthesis. Consequently, it is not surprising that increases in potassium availability are associated with increases in plant height, leaf size, and weight, stomata number and size, net photosynthesis, and CO_2 assimilation.

There are many soils in North America in which potassium is adequate for all legume needs, but this element is so important for legume growth that it should be added, on an annual basis, where there is any danger of a deficiency.

Sulfur. This substance is one of the major limiting factors for the production of legume herbage on the North American continent. From northern Saskatchewan in Canada to southern California in the United States, some of the most striking fertilizer effects are achieved by applications of sulfur to legumes. Ammonium sulfate, sodium sulfate, potassium sulfate, magnesium sulfate, gypsum, and elemental sulfur have all been used successfully where sulfur is deficient. On sulfur-deficient soils, applications of this substance increase the nitrogen, as well as the sulfur, content of the herbage. They also improve survival and resistance to winterkill in forage legumes.

TABLE 13-5
Nine-year average alfalfa yield and percentage stand survival at the end of the trial

Potassium rate		9-year average dry-matter yield		Alfalfa stand at end of trial
kg/ha	lb/acre	kg/ha	lb/acre	%
0	0	4,100	3,649	0
47	42	6,100	5,429	0
93	83	7,700	6,853	> 20
187	166	8,900	7,921	> 50
280	249	9,500	8,455	< 50
376	334	9,800	8,722	70

(Source: Agron. J., 1965)

Grass-Legume Mixtures

The total response to fertilizers of a grass–legume mixture differs from the response of individual species in a pure stand. Components of a mixture respond differently to a particular nutrient and also interact differently with each other. Consequently, management practices, including fertilizer treatments, can cause a sward to become legume dominant or grass dominant. The factors which determine the dominant type can be numerous, but the most important factor is competition for light. For the grasses, plant density, height, and nitrogen supply are all correlated. Increased nitrogen applications will increase grass yields; this increase gives a higher leaf area of grass to overshadow the lower-growing legume canopy and reduce the light intensity at the legume leaf surface. This, in turn, reduces photosynthesis and slows legume growth.

Competition for potassium also exists in a grass–legume mixture and influences growth rates, increasing that of the grass, again resulting in a decrease in light intensity for the legume. The cation exchange capacity of dicotyledonous plant roots is higher than that for monocotyledonous plants. Thus, while legume roots absorb large amounts of calcium, they may be unable to successfully compete with grasses for potassium.

Lime. Heavy applications of nitrogen fertilizers, which have a residual acid effect, can create strongly acid soil conditions in a matter of a few years. In the top soil, this may be overcome by applications of lime. The problem in the subsoil is far more serious and has no short-term solution. In humid regions, subsoil acidity will limit effective water use by forage crops. When soils are acid, the toxic effects of aluminum and manganese restrict legume growth and, to a lesser extent, grass growth. Liming decreases the chemical availability of aluminum, manganese, and iron, but increases the availability of phosphates in the soil. Plants develop finely divided and extensive root systems if aluminum levels are not toxic, thus improving root soil contact.

Liming not only affects soil nutrient availability, but it also influences soil microbiological relationships. The way in which soil acidity influences the number and activity of the microflora is not fully understood. However, it is evident that soil acidity has an important influence on nitrogen transformation. These considerations are especially important for legume crops, where three factors are involved:

1. Survival of the symbiotic bacteria in the soil.
2. General condition of the plant.
3. Effectiveness of the microorganism after the root is infected.

Strains of *Rhizobium* are variable (see Chapters 4 and 6). Some are acid tolerant; others are not. Alfalfa rhizobia are especially sensitive to pH levels below 5.0. Because populations of *Rhizobium* in the soil increase following liming, the chance encounters between the root tip and the bacteria, on which infection depends, are more likely to occur. Hence, more nodules are produced. The increase in effectiveness of the host–rhizobium relationship as pH increases may also be the outcome of enhanced molybdenum availability with rising pH. Molybdenum plays an important role in symbiotic nitrogen fixation.

As far as plant response to lime is concerned, the cool-season grasses have, surprisingly enough, received little attention. This may well be because these grasses are usually associated with legumes which require a higher soil pH. Many cool-season grasses are quite tolerant of lower pH values. Kentucky bluegrass, perennial ryegrass, and orchardgrass, for example, grow well at pH 6.5 and below if adequately fertilized. Smooth bromegrass has a higher pH requirement than other cool-season grass species.

The warm-season grasses vary widely in their tolerance of soil acidity. Coastal bermudagrass roots will withstand soils with pH values of 4.0 to 4.5. Similar results have been obtained with napiergrass, guineagrass, and pangolagrass.

Of the legumes, alfalfa is the most sensitive to soil pH, followed by sainfoin and sweetclover. The clovers, and especially white clover, are somewhat acid-tolerant, and in some cases optimum yields are obtained at pH 6.0. Red clover and crimson clover are relatively acid tolerant. Some tropical legumes such as kudzu and centro show little response to lime, but *Stylosanthes* and *Phaseolus* species give higher yields at a soil pH of 6.4 than at pH 5.2.

Plant and Animal Waste

Animal and human manure, as well as the by-products from many of the food-processing industries, contain considerable amounts of nitrogen. Although the fertilizer

TABLE 13-6A
Dry-matter yield of sudangrass and barley forage treated with dry and liquid manure and with high and low rates of irrigation in California (metric units)

Annual rate of manure on an air-dried basis (T/ha)	Dry manure		Liquid manure	
	Low irrigation	High irrigation	Low irrigation	High irrigation
	Sudangrass yield (T/ha)			
0	5.3	4.0	5.3	4.0
37	12.3	12.9	—	—
73	11.5	12.7	—	—
146	7.1	9.1	—	—
23	—	—	14.8	14.0
46	—	—	15.7	13.3
	Barley yield (T/ha)			
0	7.4	5.5	7.4	5.5
37	9.7	9.2	—	—
73	7.3	8.8	—	—
146	8.2	7.9	—	—
23	—	—	6.0	5.1
46	—	—	4.0	4.0

(Source: Pratt et al., 1976, Agron. J. 68. By permission of The American Society of Agronomy, Inc.)

industry was, in 1900, a means of disposing of wastes from slaughterhouses, such substances have now been largely replaced by chemical compounds which are used as fertilizers. Livestock wastes, once the backbone of European farming, are now regarded as a "disposal problem" rather than a valuable source of plant nutrients.

The reason for this change is that, today, the economics of handling farmyard manure are such that it is no longer competitive with the chemical fertilizers. Also, many large stockyard and poultry operations do not have cropland or pasture. As far as the economic situation is concerned, increasing energy prices, which will increase fertilizer costs, could well result in a return to the use of animal and human manure. For both types of manure, lagoons in which the wastes are decomposed by aerobic bacteria may be used to produce either liquid or solid material for redistribution on cropland.

Direct land disposal of animal waste is an excellent way to recycle manure and improve soil fertility and structure, while giving substantial forage yield increases (Table 13-6). Animal manure may also be dehydrated, bagged, and sold. The composition of fresh animal manure for different types of livestock is set out in Table 13-7. While this information indicates that animal manures are low-grade fertilizers of variable composition, the total national production of animal manure represents a vast source of unused plant nutrients. Methods must be devised which will simultaneously enable these nutrients to be used for food production and solve the present disposal problem.

TABLE 13-6B
Dry-matter yield of sudangrass and barley forage treated with dry and liquid manure and with high and low rates of irrigation in California (English units)

Annual rate of manure on an air-dried basis (tons/acre)	Dry manure		Liquid manure	
	Low irrigation	High irrigation	Low irrigation	High irrigation
	Sudangrass yield (tons/acre)			
0	2.4	1.8	5.8	1.8
16.65	5.5	5.8	—	—
32.85	5.2	5.7	—	—
65.70	3.2	4.1	—	—
10.35	—	—	6.7	6.3
20.7	—	—	7.1	6.0
	Barley yield (tons/acre)			
0	3.3	2.5	3.3	2.5
16.65	4.4	4.1	—	—
32.85	3.3	3.9	—	—
65.70	3.7	3.6	—	—
10.35	—	—	2.7	2.3
20.70	—	—	1.8	1.8

(Source: Pratt et al., 1976, Agron. J. 68. By permission of The American Society of Agronomy, Inc.)

TABLE 13-7
Total composition of liquid and solid animal manure

METRIC

Animal	% Water	Kilograms per 1,000 kg of manure		
		N	P_2O_5	K_2O
Cow	78	4.5–5.0	1.5–3.5	3.5–5.0
Horse	63	5.0–7.5	2.0–5.0	5.5–9.0
Hog	74	5.0–8.5	4.0–8.5	3.5–7.5
Sheep	63	7.5–17.0	4.0–6.0	5.5–14.0
Chicken	58	4.5–16.5	7.0–10.0	2.5–6.0

ENGLISH

Animal	% Water	Pounds per ton of manure		
		N	P_2O_5	K_2O
Cow	78	9–10	3–7	7–10
Horse	63	10–15	4–10	11–18
Hog	74	10–17	8–17	7–15
Sheep	63	13–34	8–12	11–28
Chicken	58	9–25	14–20	5–12

(Source: U. S. Jones, 1979, Fertilizers and Soil Fertility)

There is a large range of different types of waste material available from the food-processing industry. Dried blood, meat meal, process tankage, leather, fish scraps, cottonseed meal, castor pomace, cocoa shell meal, linseed meal, and compost made from all kinds of crop residues are all substances which have been used as fertilizers in the past. They could prove of value again as chemical fertilizer costs rise.

NITROGEN BALANCE IN PASTURES

An accurate assessment of nitrogen movement in the field, throughout the growing season, is difficult to determine. In grazed pastures, nitrogen may circulate from soil to herbage to livestock and back into the soil again several times during the same season. Consequently, the nitrogen cycle in Figure 13-3 should be regarded as a flow sheet of possibilities, rather than as a closed system in which nitrogen circulates. It is important that pasture managers have a generalized mental picture of the way in which nitrogen moves in their pastures and the influence that management decisions may have. The presence of a legume, or grazing instead of making hay, may profoundly influence nitrogen movement.

The atmosphere is the original source of almost all the nitrogen in the soil–plant–animal system. Nitrogen from the atmosphere may be fixed by symbiotic *Rhizobium* bacteria or by free-living organisms, mainly *Azotobacter*. The extent of symbiotic nitrogen fixation has been determined for many parts of the world. Results indicate that a well-grown legume will fix between 150 and 275 kg/ha (133 and 245 lb/acre) per

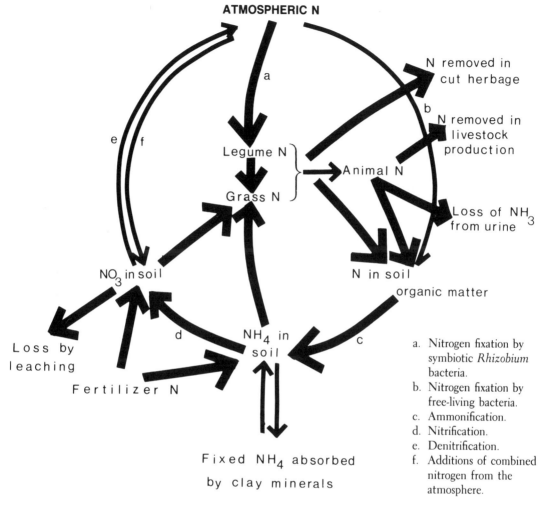

FIGURE 13-3
Nitrogen transformations in the grassland ecosystem.
(Source: D. T. Whitehead, 1970, The Role of
Nitrogen in Grassland Productivity)

year. There are a few reports of higher amounts being fixed (e.g., 600 kg/ha/yr or 534 lb/acre/yr, in New Zealand). The transfer of this nutrient material from the legume to the grass takes place mainly through the grazing animal, since 75% to 90% of the ingested nitrogen is excreted. There is also some nitrogen released into the soil from root nodules which die or are "sloughed off" from the plant and from dead leaves and plant trash which fall onto the soil surface.

The amount of nitrogen removed per hectare per year by the grazing animal is relatively small. A cow producing 4,500 l (900 gal) of milk (containing 0.64% nitrogen) per year will remove about 29 kg (64 lb) of nitrogen from the soil. The production of 360 kg of beef per year or 450 kg of lamb (containing 2.4% nitrogen) would result in the removal of 9 kg (19.8 lb) and 11 kg (24.2 lb) of nitrogen for the beef cattle, and for sheep and lambs, respectively. Thus, even at these high levels of production, the removal of nitrogenous material is not extensive, provided the animal excreta is returned to the soil, although such returns may well be in concentrated patches which require distribution. This may be contrasted with the 180 to 225 kg of nitrogen/ha/yr (160 to 200 lb/acre/yr) that might be removed in a high-yielding hay or silage crop.

The major part of the nitrogen excreted by livestock is usually found in the urine (75% to 80%). Reports which indicate a high proportion of nitrogen in the feces are due to a high nitrogen content in the diet. Some of the urine nitrogen is always lost as gaseous ammonia. A number of studies show a loss of about 6 kg/ha/hr (5 lb/acre/hr) from urine patches. During warm, sunny weather, about half the nitrogen in the urine patches is lost in this way. The nitrogen in the urine occurs mainly as urea or amino-N, and is consequently readily available for plant and microbial uptake. Nitrogen contained in feces is mostly insoluble and has to be broken down by soil microorganisms before it can be incorporated into the soil.

It is difficult to determine the amount of nitrogen which may be fixed by free-living soil microorganisms (pathway b, Figure 13-3.). Both *Clostridium* and *Azotobacter* require a readily available source of energy and reduced oxygen tension to fix atmospheric nitrogen. Both of these conditions will be more easily met under pasture, rather than with arable cropland. For pastures not receiving nitrogen fertilizer, nonsymbiotic nitrogen fixation may well be about 50 kg/ha/yr (45 lb/acre/yr). The microorganisms concerned require 500 to 2,500 kg (1,100 to 5,500 lb) of high-energy organic matter to produce this amount of nitrogen in an available form. These organisms function well at low temperatures; there are some indications that they play an important role in nitrogen fixation in northern Canada.

Pathway f in Figure 13-3 shows that small quantities of nitrate-nitrogen and ammonium-nitrogen found in the atmosphere are carried down to the soil in rainfall. Under all pastures, organic matter and the nitrogen associated with it will both accumulate and be utilized. The balance will favor accumulation in newly established pastures and, eventually, an equilibrium will be reached. This process may take a long time, some studies indicating that periods of over 100 years are frequently involved. In the grass sward, both roots and herbage, which contain nitrogen, are continually being replenished. The fact that the plant material belowground is as extensive as that aboveground is frequently overlooked. The amount of dry matter added to the soil each year from grass roots may be as much as 4,000 to 5,000 kg/ha (3,560 to 4,450 lb/acre). Under a pasture, there will be less oxygen available for the breakdown of organic matter than is the case in arable cropland, and decomposition is slow. Also, the carbon to nitrogen (C : N) ratio of organic matter will influence the rates of decomposition and immobilization. At a low C . N ratio (25 : 1 or lower), decomposition and the release of nitrogen will occur readily, while at a C : N ratio of 30 : 1 or higher, the immobilization of mineral nitrogen takes place. The C : N ratio for grass roots is usually greater than 30 : 1, while for legume roots the C : N ratio is often below 20 : 1 (pathways c and d, Figure 13-3).

While the losses of nitrogen due to leaching are considered small for most soil types when rainfall is low, high losses do occur in humid areas. They may be high where heavy dressings of chemical fertilizer are used under high rainfall conditions or with irrigation. There are also small gaseous losses of nitrogen in the form of ammonia, molecular nitrogen, and nitrous oxide from the soil.

The practical conclusions that may be drawn from these considerations of the way in which nitrogen moves in our pastures are as follows:

1. A legume will fix enough nitrogen to replace a chemical fertilizer dressing of about 200 kg/ha (178 lb/acre) per year.
2. There is a larger accumulation of nitrogen in soil organic matter under a pasture containing a legume than when no legume is present.
3. A hay crop removes a considerable amount of nitrogen (about 200 kg/ha or 178 lb/acre per year); this nutrient material must be replaced if yields are to be maintained.
4. The grazing animal removes little nitrogen and maintains soil fertility by contributing to the buildup of organic matter.

FURTHER READING

Jones, U. S., 1979, Fertilizers and Soil Fertility, Reston Publishing Co., Reston, Virginia, p. 368.

Stelly, M., 1974, Forage Fertilization, American Society of Agronomy, Madison, Wisconsin, p. 620.

Whitehead, D. C., 1970, The Role of Nitrogen in Grassland Productivity, Commonwealth Bureau of Pasture and Field Crops, Berkshire, England, p. 202.

chapter 14

The Pests of Forage Crops

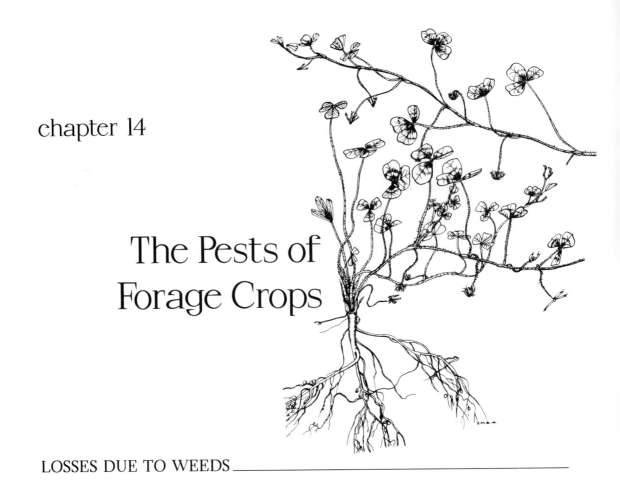

LOSSES DUE TO WEEDS

As with losses due to other pests, it is difficult to determine the extent of the crop loss that may be attributed to weed infestations. For Canada, the literature indicates an annual loss of $200 million, while the figure for the United States is $5 billion. Probably both of these estimates are conservative. Whatever the exact figure may be, the crop losses caused by weeds represent a much larger sum than the combined losses caused by insect pests and diseases.

The loss is substantial, because weeds act in a variety of different ways to reduce productivity. The main loss is the outcome of competition between the weeds and the crop for light, moisture, and soil nutrients. Other losses result from weeds harboring both insects and disease organisms. Some weeds contain toxic substances which reduce livestock productivity and may lead to animal death, while others may taint dairy products with odors or flavors, so reducing their market value. Heavy stands of noxious weeds will depreciate land values, as well.

There are also a number of nonagricultural losses for which weeds are responsible. Utility lines, highways, and railways all must be kept clear of weeds, often at great expense. There is a weed control cost in cities, where vacant lots, waste places, and lawns must be treated. Of these total losses, the loss of forage crop production represents a substantial part.

CHARACTERISTICS OF WEEDS

A weed is any plant which grows where we do not wish it to grow; it is frequently defined as "a plant out of place." It is true to say that our weed problem is one which we have created. Our grain fields, row crops, pastures, hayfields, and lawns are artificial habitats without which many of our weeds could not survive. For example, weeds in lawns may be able to survive because they are close to the ground and are not cut when the lawn is mowed. In the same way, pasture weeds may thrive because they have spines, dense hairs, or an unpleasant taste or odor, and consequently are not grazed.

Weeds frequently develop some extreme of growth habit. They are often able to grow under a wide range of climatic and soil conditions. Their seed production characteristics are frequently similar to those of crop plants with which they grow. They are also troublesome because they may produce an abundance of seeds which are the same size as the seed of one or more of our crop plants or because they have special dissemination mechanisms (e.g., barbs or tufts of hair). The seeds of numerous weed species are long lived and may remain viable in the soil for years so that the weeds persist even when further seeding is prevented. Perennial weeds, in particular, may continue to grow after the removal of their aboveground parts. The roots of such weed species are frequently spread during cultivation. Toadflax will develop shoot buds on its roots, thereby producing new surface shoots. Many weed species consist of a range of different physiological forms capable of growing in the varied conditions presented by different habitats. For this reason, many common weeds have an almost universal distribution. Weeds of European origin are, for example, now widely found in the fields of North America.

Circumstances in grasslands are somewhat different from those in arable fields. The pasture is maintained by the action of the grazing animal. Frequently, the common weeds of such areas are the climax vegetation of a forest from which the grassland was developed. Shrubs and trees are able to reestablish in grassland where grazing is inadequate or in other cases excessive. The other common type of pasture weed consists of plants which are well adapted to grazing conditions, but which give a lower yield of poorer-quality herbage than do sown species. Where overgrazing occurs, rosette plants and plants of low stature which remain ungrazed form a large part of the herbage and are regarded as weeds because of their poor yield.

Where the rotation consists of both row and grassland crops, weeds of arable land are found in forage stands during the establishment year. However, most of these weeds are annuals and do not survive after the first year. In a well-established pasture, there is little opportunity for weed seedling establishment. The weeds which are frequently associated with some common forage species are listed in Table 14-1.

Weeds are usually classified in relation to their duration and type of vegetative growth. These are important considerations in relation to the control methods which may be employed. Consequently, weeds are said to be annuals, biennials, perennials without special types of vegetative propagation, or perennials with special vegetative propagation features.

Annual weeds are usually killed by cultivation before the forage crops are established. In this case, infestation problems arise from the rapid growth and large seed

TABLE 14-1
Weedy species, in order of frequency of occurrence, commonly associated with certain forage crops

ORCHARDGRASS
    Rumex crispus, curled dock
    Silene cucubalus, bladder campion

PERENNIAL RYEGRASS
    Rumex crispus, curled dock
    Ranunculus repens, creeping buttercup
    Plantago lanceolata, narrow-leaved plantain
    Rumex acetosa, garden sorrel
    Stellaria media, chickweed

RED CLOVER
    Plantago lanceolata, narrow-leaved plantain
    Rumex crispus, curled dock
    Plantago major, broad-leaved plantain
    Chenopodium album, Lambsquarters

SAINFOIN
    Bromus pp.
    Rumex crispus, curled dock

TIMOTHY
    Plantago major, broad-leaved plantain
    Rumex acetosa, garden sorrel
    Cerastium vulgatum, mouse-ear chickweed
    Rumex crispus, curled dock

WHITE CLOVER
    Rumex acetosa, garden sorrel
    Plantago lancelata, narrow-leaved plantain
    Plantago major, broad-leaved plantain
    Rumex crispus, curled dock

(Source: Collected from various Agriculture Canada and Alberta Agriculture publications)

production by the weeds. It is possible, but not economically feasible on a field scale, to kill annual weeds at the dormant seed stage. Therefore, the plants must be killed after germination. Biennial weeds and perennial weeds with no special vegetative propagation feature are readily killed by cultivation before seeding or by mowing at a height that will not affect the forage in the establishment year. Perennial weeds with creeping or underground stems or with roots that give rise to adventitious roots are the most difficult weeds to control.

POISONOUS WEEDS

On the North American continent, there are over 100 species of plants which are poisonous to livestock. These plants are dangerous because they contain one or more of the antiquality substances discussed in Chapter 10 (e.g., alkaloids and glucosides) or because of the presence of resinoids, oxalic acid, trematol, or selenium. The most common poisonous plants are as follows:

Bracken fern	*Pteridium* spp.
White snakeroot	*Eupatorium rugosum*
Halogeton	*Halogeton glomeratus*
Seaside arrowgrass	*Triglochin maritima* L.
Death camas	*Zygadenus gramineus* Rydb.
Lady's thumb	*Polygonum* spp.
Greasewood	*Sarcobatus vermiculatus* (Hook.) Torr.
Low larkspur	*Delphinium bicolor* Nutt.
Tall larkspur	*Delphinium scopulorum* Gray
Chokecherry	*Prunus* spp.
Lupines	*Lupinus argenteus* Pursh
Milk vetches	*Astragalus pectinatus* (Hook.) Dougl.
	Astragalus bisulcatus (Hook.) A. Gray
Timber milk vetch	*Astragalus serotinus* A. Gray
Golden bean	*Thermopsis rhombifolia* (Nutt.) Richards
Locoweed	*Oxytropis* spp.
Water hemlock	*Cicuta douglasii* (DC.) Coult. & Rose
Baneberry	*Actaea rubra* (Ait.) Willd.
Poison ivy	*Rhus radicans* L.

Many poisonous weeds grow in wet or swampy locations. Seaside arrowgrass, in which prussic acid is the poisonous substance, grows on wet saline soils. Lady's thumb is found in marshes and nonalkaline sloughs. This palatable, high-protein plant, which causes photosensitization or yellows in sheep, is associated with a loss of weight and condition. Animals usually recover rapidly from the disease if they are kept out of direct sunlight and given a new diet. Water hemlock, as the name suggests, also grows in wet places, and may produce green herbage early in the spring before other plants start to grow. The back legs of cattle grazing water hemlock become paralyzed; the animals stagger, have dilated and glassy eyes, and finally become comatose and die. The disease symptoms for sheep feeding on water hemlock are similar to those for cattle: breathing is labored and the animal is unable to stand.

Several other poisonous plants are dangerous because, like water hemlock, they start growth early. Death camas and the larkspurs are examples of plants of this type. Losses caused by death camas usually occur on overgrazed pastures, where early growth from other species is poor. The poisonous larkspur species are short lived, and here, too, losses usually occur in the spring or early summer.

The poisonous weeds found in the warm, humid areas in the southeastern United States are larkspur, crotalaria, horse nettle, bladderpod, black nightshade, pokeweed, death camas, Indian hemp, white snakeroot, and milkweed. High temperatures and humidity in this region give rapid growth. Consequently, species like sorghum, sudangrass, and johnsongrass are more likely to develop dangerous amounts of prussic acid than when grown in other areas. Canada and Russian thistle, goldenrod, lambsquarters, pigweed, and smartweed will, for the same reasons, accumulate high levels of nitrates which can poison livestock.

Good pasture management practices, combined with the spot treatment of dangerous weeds with herbicides, will prevent many of the animal losses caused by poisonous plants. A dense, vigorous sward excludes poisonous plants and other weeds. For example, overgrazing is frequently accompanied by an increase in the number of death camas plants in the pasture. Fencing or draining swampy areas will reduce or exclude poisonous plant populations. Losses from poisonous plants are more common during a drought when animals will eat all the available herbage; supplementary feeding at such times is advisable. Early grazing, before the grasses are plentiful, should be avoided.

When an animal dies from a plant poison, it should be assumed that other animals in the herd have also been affected, even if there are no visible symptoms. If possible, the herd should be moved to another field until the cause of the death is eradicated. It is important to know the poisonous plants of your area and to know which parts of the plants are poisonous.

Ergot, which attacks the flowering heads of grasses and cereals (especially rye, dallisgrass, and wildrye), can be fatal to all classes of stock. The ergot fungus can cause abortion in cattle and horses, as well as blindness, diarrhea, ulcers on the lips, and gangrenous legs and hoofs. When the disease occurs, animals should be moved to another pasture, and the pasture in which ergots are present should be mowed to remove the grass heads.

WEED-CONTROL METHODS

Fortunately, the diverse and varied nature of our weed species is matched with an extensive number of control methods. These may be divided into four groups:

1. Direct destruction of the weeds, where cultivation or mechanical control

methods are used in such a way as to damage the plant by cutting or burying it.
2. Ecological methods, in which the plant is attacked indirectly by altering the environment in such a way as to make conditions less favorable to the weed.
3. Biological control, where a disease or an insect pest which will attack the plant is introduced.
4. Chemical control, which involves the use of a herbicide that is selective or nonselective in action or that is one of a large group of growth-regulating substances.

Direct Destruction

The main principle involved in this method, whether it is applied to small annual weeds or to clearing bush, is *dessication*. The weed is cut through at, or a little below, ground level and left on the surface to dry. The most important key to success is to select dry weather to aid in dessication. Seedbeds are frequently prepared during dry periods in the fall. To improve the kill and ease the work involved in areas which have cold winters, bush is often cleared at a time when parts of the plants and the soil are both frozen. Also included in this type of control is the procedure of burying the plant by plowing or cultivating in other ways. This is effective with annuals and some perennials. Mulching will have the same effect as burying; for this purpose, straw, paper, or black plastic material may be used. In general, the perennial nature of most of our forage crops limits the use of all methods of this type except at the time of establishment.

Ecological Control

For forage crops, providing satisfactory drainage frequently represents the best way to change the environment to one in which the crop, rather than the weed, is favored. There are many weedy species that can withstand prolonged flooding, while there are no forage legumes and few forage grasses that will withstand flooding for longer than 3 weeks. Adequate soil fertility levels and the application of lime to ensure a near neutral soil reaction will also favor the forage species (especially in humid areas), thereby retarding weed growth. Burning a pasture shortly after what is expected to be the last "killing" frost in the spring can control winter weeds in stands of bermudagrass. It will also remove dead grass and so hasten spring growth.

Biological Control

This term could be used to include all cases of weed control by another organism. Thus, competition from other plants and grazing by livestock could be regarded as examples of biological control. It is, however, more usual to confine the use of this term to special cases of control by insect pests or diseases. The most widely known example of biological control is the eradication of prickly pear (*Opuntia* spp.) from Australia by a moth, *Cactoblastis cactorum*. Prickly pear, which was introduced from America, became a serious weed in Australia. The moth, whose larvae feed on the prickly pear, was introduced from Argentina. It multiplied rapidly and virtually killed out the prickly pear. The spread of a disease called myxomatosis among wild rabbits in Britain substantially reduced the population of these animals and is a further example of biological control.

Chemical Control

The value of a chemical for weed control depends on the amount required, its cost, its persistence, and its selectivity. Herbicides are required for use on ground which is to be permanently bare (e.g., railway lines). Such substances should be nonselective, killing all plants, and should remain toxic for as long as possible. For temporary bare land, such as fallowed land, a quick-acting, nonselective, short-lived substance is required. Selectivity becomes important when the herbicide is used in a growing crop. For forage crops, herbicides which will control weeds in grasses, in legumes, or in a grass–legume mixture are required.

There are many methods of applying chemical substances. Nonselective herbicides are frequently applied as dry granules or as a coarse spray on the ground. Volatile substances like carbon bisulfide are injected into the soil. Selective herbicides are applied to the plant as dusts or, more usually, as sprays. Low-volume sprays (70 to 110 l/ha or 6.3 to 10 gal/acre) are widely used, since they call for small spray tanks and less water.

Nonselective Herbicides

The most long-lived general weed killers are arsenic compounds, which are little used because they are poisonous to people and animals. *Borax* and other *boron* compounds, organic substances, substituted ureas, together with *chlorates*, are useful and widely used herbicides for permanently bare land. Mineral oils are used as nonselective contact herbicides to kill a wide range of plants. The quaternary ammonium compounds are especially useful on land which is being fallowed. These substances can be used as a spray immediately before seeding, since they are both inactivated on contact with the soil.

Selective Herbicides

There are no chemicals that are "selective" at all application rates. However, there are many substances which may be applied at a rate high enough to kill weeds without damaging forage crops. A herbicide is said to be "selective" when it has a differential effect on the weed and crop plants. The smaller the difference between the two types of plants, the more difficult it becomes to produce this differential effect. It is possible to control dicotyledonous weeds in a cereal or grass crop. It is also possible to control monocotyledonous weeds in alfalfa. For most forage swards, the use of herbicides is complicated by the diversity found in both the forage and the weed species. Often there are broad-leafed species which should be killed growing beside legume species that must be retained in the sward. Legumes, the most sensitive plants in the pasture, have different degrees of tolerance to herbicides. The kind and rate of herbicide which can be tolerated by the most sensitive legume species usually determines the type of spray program used. This may well restrict the weed control that may be achieved.

The most effective substances which have these selective properties are *growth-regulator* or *hormone* weed killers. These are substituted phenoxyacetic acids, of which the most widely used are the following:

1. 2-Methyl-4-chlorophenoxy acetic acid (MCPA).
2. (2,4-Dichlorophenoxy acetic acid (2,4-D or DCPA).
3. (2,4,5-trichlorophenoxy acetic acid (2,4,5-T). This substance is banned from use in the United States.

MCPA and 2,4-D are usually used in the form of their sodium or amine salts, since these are soluble in water. 2,4,5-T is used as an ester. MCPB [4-(2-methyl-4-chlorophenoxy)-butyric acid] is not broken down in some plants and is thus inactive. Where a weed has the appropriate β-oxidation enzyme, it will convert MCPB to MCPA, which then has the normal toxic effect. The presence or absence of this enzyme thus introduces an additional type of herbicide selectivity.

Low-volume spraying is especially suitable for hormone weed killers. These substances are absorbed into the plant and translocated to all parts. They are most effective when the plant is actively growing and can be used on perennials. Even if the plants are not killed, these substances will distort the inflorescence and decrease fertility. All the perennial grasses can withstand the three hormone weed killers listed above in doses that will kill broad-leaved weeds.

ChloroIPC is very toxic to grasses and will selectively kill many dicotyledonous weeds in legumes. The treatment should be given early in the season when the legumes are dormant. 2,4-DB or a mixture of MCPB and MCPA is also effective in mature legume stands. 2,4-DB (Embutox E) may also be used in seedling stands of alfalfa and birdsfoot trefoil. While reducing weed infestations, some of these substances adversely affect the digestibility of forage legumes (Table 14-2).

TABLE 14-2
Digestion coefficients for herbicide treated and untreated alfalfa hay, Madison, Wisconsin

	Alfalfa hay	
Digestion coefficients (%)	Treated	Untreated
Dry matter	63	66
Crude protein	69	76
Acid-detergent fiber	40	41

(Source: Temme et al., 1979, Agron. J. 71. By permission of The American Society of Agronomy, Inc.)

AVOIDING INFESTATIONS

There are a number of practices, which may be classified as "good farming" methods, which will substantially reduce weed infestations in forage herbage. First and foremost is the purchase or use of *clean seed* for establishment. Inadequately cleaned forage seed may contain 1% to 5% weed seeds, and even this relatively small number will establish enough weed plants to start a major infestation. Next, the importance of a well-prepared, weed-free seedbed cannot be overemphasized. Also, timely seeding will place the forage crop at the greatest advantage in relation to the weeds. Companion crops, which were discussed in Chapter 6, can be used to suppress weeds or to inhibit their germination.

INSECT PESTS

The numbers and types of insects that live and feed on forage plants are vast. Not *all* of these are injurious, but those which are can compete successfully with grazing animals when conditions favor their multiplication. Grasshoppers, locusts, and worms can destroy crops of grass, while aphids can eliminate alfalfa and clover stands. The beneficial insects are those which act as pollinators for legume flowers and the parasites and predators of insects that damage forage crops. The basic problem of insect control is to kill the destructive insects and encourage those which are helpful.

The magnitude of the loss due to insects is very difficult to determine, but it is usually regarded as being enormous. Yields are reduced, stands are lost prematurely, and forage quality is decreased. No less than 15% of our forage production is lost in this way every year.

Types of Insects Attacking Forage

Insects will attack all parts of a forage plant. Flowers, flower buds, and seed may be destroyed; roots, rhyzomes, stem, and leaves can all be eaten by one insect or another. For agricultural purposes, insects are best classified in relation to the parts of the forage plant they attack.

Foliage Feeders. This group, members of which will attack both grasses and legumes, contains by far the largest number of insect species. Locusts, grasshoppers, caterpillars, and crickets are widely prevalent pests of pasture and hay crops. The migratory grasshopper *(Melanoplus sanguinipes)* is found over most of the North American continent. Grasshoppers deposit their eggs in pods in the soil, and the nymphs hatch in the spring. The migratory grasshopper passes through two or three generations per year in the southern United States.

There are over 60 species of cutworms and armyworms which feed on forage crops. A few of these cutworms live underground, but most bore inside the stem of their host plants. Armyworms are cutworms that are found in large numbers and which move on foot in search of food. Caterpillar populations may also be large. For example, the alfalfa caterpillar *(Colias eurytheme)* is found widely in the United States and Canada and can be a serious pest in the Southwest. The forage looper *(Caenurgina erechtea)* is normally a minor forage pest over much of the central area of North America, but in some years it causes serious damage. Webworms attack a wide range of plants and can heavily defoliate forage crops. The species of webworms concerned are *Loxostege commixtalis, L. similalis,* and *L. sticticalis.*

Weevils cause considerable damage in the southern United States. Attacks by the alfalfa weevils *(Hypera postica)* are estimated to result in an annual $57 million crop loss. Most of this damage is the result of larvae feeding on the plant tips and leaves as they open. The adult weevils, which also feed on alfalfa, do little damage. This pest spread rapidly when it was introduced to North America from Europe early in this century.

There are several species of ants which cut off foliage or seed from plants, including grasses and legumes. The Texas leaf-cutting ant does not eat such material, but will use it to grow "fungus gardens." The ant then feeds on the fungus. Yields of forage in areas around the ant's nest are badly reduced. There are 20 or more species of blister beetles which feed on forage crops. *Epicauta murina* is one of the most widely encountered. Leaf miners have small larvae which live and feed between the upper and lower leaf surface. They can cause extensive damage in legume crops.

Sap-sucking Insects. Aphids, the nymphs of the meadow spittle bug and various leafhoppers, and the three-cornered alfalfa hopper, all suck sap from the leaves and stems of forage plants. The most serious threat to forage production comes from the aphids. The pea aphid, *Acyrthosiphon pisum,* is a widely distributed and serious pest of alfalfa. The spotted alfalfa aphid, *Therioaphis maculata,* is spreading rapidly in North America. In both species, the adult aphids and the nymphs suck sap from the plants. In the warmer parts of the United States, they reproduce throughout the year. Heavy infestations can reduce the stand, prevent flowering and seed production, and reduce both yield and quality.

The meadow spittlebug *(Philaenus spumarius)* produces on forage plants a mass of white froth in which the nymphs live. The nymphs stunt the plants and reduce forage yield and quality. The adult, which resembles a leafhopper, feeds on a number of different species, but normally lays its eggs on legumes, frequently alfalfa. *Empoasca fabae,* the potato leafhopper, attacks forages in the eastern part of the United States. Both adults and nymphs pierce the leaves and stems and suck the plant juices. The plants are stunted and the translocation of carbohydrates is reduced.

Chinch bugs are small, black sucking insects which feed only on members of the grass family. They are pests of corn and small grains, but also attack forage crops. They will swarm in the hot summer months on corn, sorghum, sudangrass, and millet, killing plants in these crops. They are abundant in the central and eastern states. The lygus bug is a sap-sucking insect which is frequently found in alfalfa fields in the southwest of the United States. There are many of these species in various parts of North America, corresponding to variations in habitats and host plants. Mites are members of the class *Arachnida* and, like ticks and spiders, have eight legs. They are normally found on the undersurface of the leaves of many legume crops in areas with warm summer climates.

Root Feeders. Fewer insects attack the plant's root system than the aerial parts. There are a number of species of white grubs which attack the roots of grasses in pastures and lawns in the eastern and central United States and Canada. Permanent bluegrass pastures are attacked in this way and may be badly damaged. The adult is a large brown or black beetle called a "may beetle" or "June bug." They live mainly in trees, feeding on the foliage, and lay their eggs in grasslands.

The clover root curculio *(Sitona hispidula)* has larvae which feed on young, but fibrous, forage plant roots. The damage caused by the larvae will permit the entry of other organisms, frequently fungi, which cause plant disorders. The white-fringed beetles (*Graphognathus* spp.) have larvae which feed on the roots of legumes and many other species. They occur in the southwest of the United States and do not normally cause extensive damage.

Insects That Reduce Seed Production. Although any insect that attacks a plant may reduce seed production, blister beetles, grasshoppers, pea aphids, and mites are especially likely to influence the numbers and size of the seed. Legume crops are more frequently attacked than grasses. There are some species of *Lygus* (e.g., *L. hesperus*, *L. lineolaris*, and *L. elisus*) which feed almost exclusively on seeds and flowers, causing seeds to shrivel and flowers to drop. Several other plant bugs and stink bugs damage legume flowers in the same way. All these insects are able to insert their mouthparts through the pods into the seed and suck out the contents. Seed chalcids have larvae which live inside a seed. These larvae eat the seed as they develop. The common field cricket will eat the developing seed pods at the milk stage.

INSECT CONTROL

Forage crops cover large areas and are often subject to extensive, rather than intensive, use. An important feature of their value to the farmer is that the animal feed they provide is readily and cheaply obtained. To preserve this feature, it is essential that the cost of insect control (like weed and disease control) be as low as possible. The most expensive control system is the use of chemicals to kill the insects attacking the crop. To reduce costs, this approach should be used only in conjunction with one of the three other control methods. These are (1) the use of insect resistant cultivars, (2) biological control, and (3) cultural control.

Cultural Control

The intention of cultural control is to alter normal farming practices slightly to adversely affect the insect pest. Frequently, this may be achieved by advancing or delaying the harvest. Alfalfa pests provide a number of examples of insect control of this type. For example, the eggs of the potato leafhopper may be removed from the field with the forage if harvest is delayed. Early harvesting also reduces insect populations by exposing them to high temperatures when the crop is cut for hay. Caterpillars may be controlled in this way, over a wide area, if a community acts together. The use of crop rotations to control insect pests has long been practiced. The cultural control of insect pests must always have a place in agriculture, though it may be used together with other methods.

Resistant Cultivars

The primary advantage of resistant cultivars is that they provide a long-lasting solution at little or no cost to the grower. Some examples of insect-resistant cultivars of alfalfa are those which will withstand attacks by spotted and pea aphids, while other cultivars are tolerant of alfalfa weevils, potato leafhoppers, and meadow spittlebugs attacks.

Biological Control

The use of biological control of weeds was discussed earlier in this chapter. The same principles apply to the biological control of insects. For example, parasites of the alfalfa weevil have given successful control when released in both the western and eastern United States. There have also been attempts to establish parasites and predators (such as the lady beetle, *Hippodamia convergens*) to control both spotted alfalfa aphids and pea aphids. The parasites and predators, once established, maintain themselves and spread naturally. Populations of the lady beetles are, however, slow to build up, and the aphids do considerable damage before they are brought under control. The application of biological control methods calls for a careful and extensive study of the life histories of the insects which attack forage crops, as well as those of their parasites and predators.

Chemical Control

There are many types of chemical substances which will give very effective insect protection to forage crops. The chlorinated hydrocarbons, for example, give excellent control of many insect pests. However, these substances leave residues on the crop which are found in the products of animals that feed on the forage. Organophosphates (like parathion and malathion) are also effective, but leave no harmful residue if properly applied. Carbamate carbaryl can also be used. These substances are available as emulsifiable concentrates, wettable powders, granules, dusts, and oil solutions. It is important that the correct insecticide be applied at the correct time and in the correct way. Local advice should be obtained as to what is appropriate, since new and improved insecticides are continually being developed. It is also important to remember that insecticides which are not properly applied are injurious to humans, animals, plants, fish, beneficial insects, and other wildlife. Insecticides should be used in conjunction with the other control measures discussed here, and not be regarded as a quick and easy answer to all insect pest problems.

FORAGE CROP DISEASES

Forage crops are hosts to bacteria, fungi, viruses, mycoplasma, nematodes, and a few parasitic higher plants, all of which cause diseases and crop loss. Attacks by fungi are most common, but each type of causal organism is responsible for several economically important diseases. Every forage crop species is attacked by many different diseases, although not all of them are serious. There are 45 known diseases of Kentucky bluegrass, 35 of timothy, and over 30 of orchardgrass, one or more of which attack all parts of these plants. Roots are attacked by soil-living fungi and nematodes. Leaf and stem disorders such as blights, mildews, leaf spots, rusts, and smuts are caused by fungi, bacteria, and viruses. Ergots, head and kernel smuts, and seed galls (which attack seeds and flowers) are produced by fungi and nematodes.

As well as depleting yield, plant diseases usually reduce forage quality. The more readily digestible carbohydrate substances are used by the parasites, thereby reducing the digestibility of the forage. Fewer photosynthates are available for protein synthesis in the diseased plant, causing crude protein levels to be low.

DISEASES OF LEGUMES

Bacterial Diseases

Bacterial wilt *(Corynebacterium insidiosum)*, bacterial stem blight *(Pseudomonas medicaginis)*, and bacterial leaf spot *(Xanthomonas alfalfae)* commonly attack alfalfa. Bacterial stem blight and bacterial leaf spot are of minor economic importance. Bacterial wilt, whose symptoms include stunting and yellowing of the entire plant, is widespread and can kill susceptible crops in from 2 to 4 years. Diseased plants normally wilt during the summer, because the plant vessels are plugged with bacterial growth. The disease may be seed-borne and is also carried on plant debris. Cultivars resistant to the disease are commercially available. High soil fertility levels enable plants to withstand attacks.

Fungus Diseases

The most conspicuous and numerous fungus diseases of legumes are the leaf spots. These are found, to some degree, in almost every legume crop. For example, common leaf spot *(Pseudopeziza medicaginis)*, a destructive disease of alfalfa, occurs wherever the crop is grown. Early harvesting will help to prevent the spread of this disease, and resistant cultivars are available. Stemphylium leaf spot reduces forage production during wet periods. The causal organism is *Stemphylium botryosum*.

Powdery mildew *(Erysiphe polygoni)*, which attacks red and alsike clover, causes a substantial loss in hay yield and quality every year. A white mat of mycelia covers the clover leaves, reducing photosynthesis and increasing transporation. The leaves actually die and plant growth is stunted.

The rusts, *Uromyces striatus* and *Uromyces trifolii*, attack alfalfa and common clovers, respectively. For alfalfa, the reduction in seed yield can be extensive, and in late hay crops, leaves are lost and the quality is reduced. Rust is sometimes severe on white, alsike, and red clover.

There are two common anthracnose diseases *(Colletotrichum trifolii* and *Kabatiella caulivora)* which reduce the forage yield of red clover. Anthracnose is also found on alfalfa, and in the mid-eastern United States it is the main cause of "summer decline." The plants develop sunken lesions on the stems. These lesions, which tend to coalesce at or near the soil level, girdle and kill the stem. There are some alfalfa cultivars with resistance to this disease.

Stem and root rots of forage legumes are caused by *Sclerotinia trifoliorum*. This fungus is active in late winter and early spring, when it will attack the stem bases and upper taproots, killing some stems. Serious outbreaks of sclerotinia have occurred in crimson and white clover, sweetclover, and alfalfa. Phytophthora root rot *(Phytophthora megasperma)* is found in alfalfa growing on poorly drained or overirrigated soils. It will attack both seedlings and mature plants, destroying the taproot. If the disease progresses to the crown, the plant dies. A number of *Fusarium* species damage legume roots, usually attacking the plants after an injury. The disease spreads slowly in mature stands. For seedlings, however, plant death may rapidly follow infection. Pythium root rot (caused by a number of *Pythium* species) commonly results in postemergence damping-off of young legume seedlings.

Virus and Mycoplasma Diseases

Viruses cause dwarfing, witches broom, and mosaic diseases in legumes. Mycoplasma are bacterialike organisms, but, unlike bacteria, are small enough to be filterable. They cause yellows and dwarf-type diseases.

Virus mosaic disease is characterised by a yellow and green mottling of the leaves. The plants are frequently stunted. This disease is widespread, but the losses do not appear to be of major importance. It is also found on red, alsike, white, and crimson clover, sweet clover, and alfalfa. Alfalfa dwarf disease occurs most frequently in southern California and has symptoms similar to those of bacterial wilt. The diseased plants are normal in color or slightly darker green, and not yellow, as when bacterial wilt is present. Witches broom or "bunchy top" is found in Washington state.

Nematodes attack the roots of most legumes and some grasses. Because they form swellings on the roots, they are known as *root-knot* nematodes. Legume stems are also attacked. The stem nematode *(Ditylenchus dipsaci)* is more destructive than are the root-knot nematodes *(Meloidogyne hapla* and *M. javanica)*. Stem nematodes attack stem bases; crown buds are frequently infected and fall off the plant. The root-knot nematodes cause dwarfing of the plant and extensive branching of the root. The grass seed nematode produces galls which replace the normal seed of red fescues and bentgrasses.

LEGUME DISEASES IN THE SOUTHEASTERN UNITED STATES

In the warm, humid climate of the southeastern United States, legume diseases on all types of forage (winter and summer annuals and perennials) are very likely to develop. Of the winter annuals, field peas are attacked by *Ascochyta pinodella* and *Mycosphaerella pinodes,* which produce a condition known as *black stem.* Leaf blotch *(Septoria pisi)* is common on field peas, turning leaves yellow and then brown. Powdery mildew *(Erysiphe polygoni)* also contributes to field pea losses. Vetches are attacked by anthracnose, but resistant cultivars are available. Crimson clover is subject to stem and crown rots *(Sclerotinia trifoliorum),* which develop and spread rapidly in cool, wet weather.

Turning next to the summer annuals, dodder *(Cuscuta arvensis)* is a higher plant parasitic on lespedeza. Once dodder seed is present in the soil, it is able to persist for many years. Lespedeza is attacked by powdery mildew *(Microsphaera diffusa)* and root-knot nematodes. Cowpeas are subject to wilt *(Fusarium oxysporum),* but some resistant cultivars exist. The most serious disease of kudzu (a perennial) is the root-rot fungus *Rhizoctonia solani.* The leaflets of kudzu are commonly affected by halo blight *(Pseudomonas phaseolicola).* Defoliation occurs only when spots are numerous. The plant is able to produce new leaves.

DISEASES OF GRASSES

The organisms which attack grasses are many and varied. The diseases which cause serious damage to the most widely grown grass species and the location where they occur are given in Table 14-3.

Warm-Season Grasses

Species of the genus *Helminthosporium* attack bermudagrass to cause leaf eyespot *(H. giganteum)* and leaf bleaching and withering *(H. cynodontis).* Carpetgrass leaves are also spotted by species of *Helminthosporium* and by *Curvulacia.* Dallisgrass is attacked by anthracnose *(Colletotrichum graminicola)* and by *Stagonospora paspali,* both of which produce leaf spots. Johnsongrass and sudangrass frequently suffer severe leaf injury from numerous diseases. During the cool season, tall fescue suffers from leaf scald caused by *Rhizoctonia solani,* net blotch *(H. dictyoides),* and leaf spot *(Cercospora festucae).*

The warm-season annual grasses are attacked by bacterial stripe *(Pseudomonas andropogoni).* This disease is particularly destructive when found on sudangrass. It also occurs in johnsongrass and sorghum. The lesions, which are up to 1 ft (0.33 m) long, vary in color from purple to brown. Bacterial exudates frequently form a crust over the lesions. Bacterial streak *(Xanthomonas holcicola),* the leaf-blight fungus *(H. turcicum),* target spot *(H. sorghicola), H. rostratum,* and anthracnose also attack sudangrass. The bacterial spot *(Pseudomonas syringae)* is not common, but in some years it is found

TABLE 14-3
Serious diseases of important forage grasses and their geographic distribution

Grass species	Disease	Location
Bermudagrass	Rhizoctonia rot	Southern states
Bluestems	Seedling blight Seed rots	Great Plains and prairies
Bromegrasses	Seedling blight Seed rots Snow molds Root rot	Great Plains and prairies Northern areas General
Kentucky bluegrass	Leaf spot Brown patch Root rot	Northern areas General
Orchardgrass	Root necrosis Seed rot Seedling blight	General
Ryegrasses (Lolium)	Root rot Rhizoctonia rot Seed rot	General
Timothy	Seedling blight Seed rot Root necrosis	General
Wheatgrasses	Seedling blight Root rot Crown rot	Western states and provinces
Wild ryegrasses	Seedling blight Root rot Crown rot	Western states and provinces

(Source: U.S. Department of Agriculture Year Book, 1953)

abundantly on sudangrass, johnsongrass, pearl millet, foxtail millet, sorghum, and corn. Fortunately, the level of infection from the diseases that attack sudangrass and related species may be substantially reduced by seed treatment and crop rotation.

Cool-Season Grasses

Many of the diseases which attack the warm-season grasses are also found among the cool-season species. *Helminthosporium bromi* (brown spot) produces small, dark-brown spots on the leaves of bromegrass which develop first in the spring. Fruiting bodies of *H. bromi,* on dead leaves from the previous year, discharge ascospores which are carried by the wind to infect new growth. *Rhynchosporium secalis* causes leaf scald on smooth bromegrass, and a fungus from the same genus also attacks orchardgrass. The symptoms are similar on these two hosts. Water-soaked spots first appear and then enlarge to become scaldlike blotches with brown margins. If conditions are favorable, the disease becomes more destructive as the season advances.

Leaf spot *(Selenophoma bromigena)* attacks several members of the genus *Bromus* in the central and western United States and Canada. A small brown fleck appears on the leaf in the spring. As the season progresses, the disease spreads to the stem. The bacterial diseases *Pseudomonas coronafaciens,* brown stripe *(Scolecotrichum graminis),* and septoria leaf spot *(Septoria bromi)* also affect smooth bromegrass.

Orchardgrass is affected by many disease organisms. Brown stripe *(Scolecotrichum grammis),* scald *(Rhynchosporium orthosporum),* and the rusts occur in most areas where orchardgrass is grown. *Mastigosporium rubricosum,* which attacks orchardgrass and many other grasses, is most common during the spring and autumn. The lesions it forms are dark purple or brown flecks, like those produced by the fungus *Stagonospora maculata,* which is found mainly in the northeastern parts of North America.

Timothy is attacked by stem rusts, stripe smut, scolecotrichum, brown stripe, eyespot, and bacterial stripe. *Heterosporium phlei,* the cause of eyespot, is widespread, occurring on other grasses as well as timothy. *Xanthomonas translucens* is the organism causing bacterial stripe, and it too will affect many grasses and cereals.

Kentucky bluegrass suffers from a leaf spot caused by the fungus *Helminthosporium vagans,* which also attacks the crown. The leaf spot may be small, frequently only the size of a pinhead, or may cover the leaf blade. Infection during the growing season occurs from spores produced on older lesions.

Root and crown rots are the most destructive group of parasites of the grass family. Most of these organisms need plenty of moisture and survive for part of their life in the humus fraction of the soil. Such organisms cause preemergence rots, killing plants or seeds as they start to grow or germinate. The most common cause of seed rots are *Pythium debaryanum, P. ultimum, Fusarium culmorum,* and *Rhizoctonia solani.* Two of these organisms *(R. solani* and *P. debaryanum)* also cause damping off.

DISEASE CONTROL

It is difficult to control forage crop diseases efficiently without involving a substantial expense. There is no doubt that, if forage utilization and production is to become intensive, a new integrated system of disease control is required. For the present, the modification of farming practices to avoid disease attack, in conjunction with the use of seed dressings and resistant cultivars, represents the main defenses that are available to the farmer. The development of new control systems should have high research priority. Until the results of such efforts are available, the use of resistant crops in a rotation will reduce the severity of *Phytophthora* root rot, *Sclerotinia trifolium* and other root diseases, and seedling blights. Early mowing can be used to interrupt the life cycle of disease organisms before infestations can reach serious levels. Improved soil drainage reduces incidents of wilt, and the removal (possibly by burning) of stubble and plant refuse reduces potential disease sources. Fertilizer applications and liming help indirectly, since vigorous plants are better able to withstand disease. Where they are available, resistant varieties should be used. For the grasses, seed-borne diseases may be controlled by using dusts. Chemical control is inappropriate for seed-borne legume diseases since they affect bacterial inoculants.

FURTHER READING

Franklin, C., and G. A. Mulligan, 1970, Weeds of Canada, Queen's Printer, Ottawa.

Gill, M. T., and K. C. Vear, 1969, Agricultural Botany, G. Duckworth & Co. Ltd., London, p. 637.

Hanson, C. H. ed., Alfalfa Science and Technology, American Society of Agronomy, Madison, Wisconsin.

Stefferud, A. ed., 1948, Grasses, Yearbook of Agriculture, U.S. Department of Agriculture, Washington, D.C.

Stefferud, A., ed., 1953, Plant Diseases, Yearbook of Agriculture, U.S. Department of Agriculture, Washington, D.C.

chapter 15

Seed Production from Legumes and Grasses

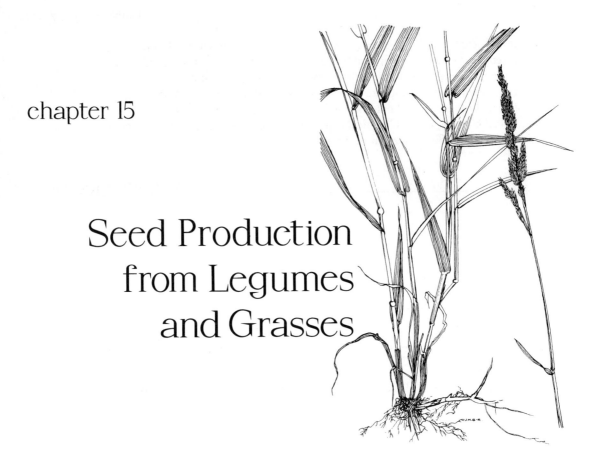

INTRODUCTION

The production of seed for forage crop establishment is an important and extensive part of the seed trade in North America. In the United States, there is annual production of nearly 300 million kg (660 million lb) of grass and legume seed, while in Canada the figure is 30 million kg (66 million lb). In the United States, the species concerned, ranked in order of magnitude of production, are annual and perennial ryegrass, alfalfa, red clover, Kentucky bluegrass, lespedeza, and timothy. The equivalent order of ranking for Canadian forage seed production is creeping red fescue, timothy, red clover, Russian wildrye, sweet clover, alsike clover, smooth bromegrass, and alfalfa. In addition, there are a large number of other species produced in both countries which are intended for use in special circumstances.

To understand the logic that underlies the organization of the forage seed trade, it is essential that we should consider the floral morphology and certain aspects of the physiology of both legumes and grasses. In the present chapter, this information will be set out first, and then the organization and methods of seed multiplication and production will be considered.

POLLINATION MECHANISMS

Nearly all our forage species are cross-pollinated, and many are almost entirely incapable of self-fertilization. The species which are normally self-fertilized are weeping lovegrass, slender wheatgrass, common and hairy vetch, lespedeza, and sudangrass. There are also some apomictic species, which include Kentucky bluegrass, buffelgrass, and dallisgrass. Two common forage grasses are dioecious (i.e., buffalograss and saltgrass), while bluestem grasses have perfect flowers and male flowers on the same plant.

All grasses have light, dry pollen and are cross-pollinated by the wind. The legumes have heavy pollen grains, which quite often have a sticky surface and so hold together in clumps, and insects, most frequently bees of various types, transfer legume pollen from flower to flower.

Grasses

Floral Morphology. The floral structure of the grasses was described in Chapter 1. Some of these features have important implications in seed production. What is sold commercially as "seed" is botanically the caryopsis, or fruit, to which are usually fused two bracts, the lemma and the palea. When flowering takes place, the anthers are extruded as the lodicules swell, thus opening the lemma and palea and exposing the feathery stigma (Figure 1-8). While the lemma and palea are open, bacteria, fungal spores, or insect eggs may become lodged on their inner surfaces. Following pollination, the ovary swells, and its outer wall fuses with the inner surface of the lemma and palea (see Chapter 1). Contaminants will thus be fused into and become an integral part of what is sold commercially as "seed." There are few insecticides, fungicides, or bactericides which can penetrate the two outer bracts and kill organisms fused to the caryopsis wall. Where such substances do exist, treatment is expensive. Consequently, it is important that grass seed production take place in districts where bacterial, fungal, and insect populations are low. The hard, cold winters in the north and in isolated mountainous regions substantially reduces the number of such organisms. Therefore, areas like the Peace River country of Alberta and parts of Oregon and Washington states are ideal for grass-seed production.

Physiological Requirements. One reason why grasses are so successful at occupying adverse environments is that they have physiological mechanisms which control, very precisely, the time of flowering. The inflorescence, while it develops, is enclosed in a leaf sheath. Each floret is protected from adverse climatic conditions by the lemma and palea, except during the brief time when the bracts are opened by the lodicules. This precise control over the time of anthesis is the outcome of the plant's response to two factors. First, day length, which is the same from year to year at any one location, provides uniformity in the timing of anthesis. For many, but not all species, low temperature, together with day length, will determine the *time* of inflorescence *initiation*. Second, the latter stages of reproductive development are controlled by temperature, light intensity, available soil moisture, and nutritional status. This second group of factors, collectively or individually, determines the *rate* of reproductive development.

In general, grasses have three conditions, or physiological stages, through which they move in response to day length and, in some cases, low temperature. As they do so, there is no *external*, visible change, though the primordia produced on the shoot apex will be reproductive (i.e., able to produce florets) rather than vegetative (i.e., producing leaves) after the third stage. These three developmental stages are (1) the juvenile stage, (2) the inductive stage, and (3) the stage of realization or initiation. Not all species have all three stages.

In the juvenile stage the plant is insensitive to factors which will subsequently induce flowering. It may be of short duration (a few days), long duration (a few weeks or many months), or nonexistent. Just what factors move the plant into the second developmental stage is uncertain, for, being a "negative" stage, the first stage is difficult to study. However, there are some indications that high light intensity, the accumulation of a minimum threshold level of carbohydrate reserve material, and a minimum threshold value for leaf area index are factors which influence this change. Examples of species which have a juvenile stage are orchardgrass, red fescue, timothy, and some cultivars of perennial ryegrass. While the plant is in the juvenile stage, it increases in size but makes no progress toward flowering.

In the inductive stage, the plant is able to respond to conditions which cause a state of ripeness to flower. These conditions are either short days (less than 12 hr), or low temperatures ($-6°$ to $10°C$; $21.2°$ to $50°F$), or some combination of the two. In fact, the second stage of development places the plant in a state of readiness to receive the stimulus which initiates flowering. This stimulus, which results in the initiation of floral primordia, is frequently long days (i.e., over 12 hr). The plant then reaches the third and final developmental stage, the stage of realization or initiation. The number of flowering buds that will produce flowering heads and the speed with which this will be accomplished is then determined by light intensity, temperature, soil moisture, and nutrients.

A number of fescue species, including creeping red fescue, pass through all three stages. Creeping red fescue has an unresponsive juvenile stage when it first germinates. In the fall, it responds to short days and low temperatures; the plant is then ripe (or ready) to receive a flowering stimulus of long days in the following spring. Smooth bromegrass has a similar response pattern, but is less precise in its requirements for induction than is creeping red fescue; induction will take place following cold-temperature treatment, but without short days.

The warm-season grasses frequently have a short-day requirement for flowering. Flowering may be delayed by low temperatures, with no flowering occurring until night temperatures exceed $12°$ to $16°C$ ($53.6°$ to $60.8°F$). High night temperatures favor flowering in the short-day grasses, but may inhibit flowering in long-day plants. Tropical grasses are able to respond to the small changes in day length found during the year near the equator or are not sensitive to day length.

TABLE 15-1
Day length in hours and days from initiation of growth to first flower production for alfalfa

Daylight hours	Days to flowering
10	47
12	38
14	28
16	24

Legumes

Physiological Responses. The precise control which day length and temperature exercise over flowering in the grasses is not found in the legumes. The day length is not, however, without effect for most temperate legumes. Flowering responses of both alfalfa and white clover change drastically from the north to the south of the American continent. Table 15-1 shows that increasing day length reduces the number of days to flowering for alfalfa. High light intensities will also increase the percentage of stems which develop reproductive buds when day length is kept constant. For most legumes, temperature, provided it is satisfactory for growth, does not influence flowering. The pollination mechanism for the forage legumes is, however, very precise and is directed toward ensuring cross-pollination.

Pollination. Different pollinating insects and different floral features play a part in pollen transfer for the various legume species. Those present in alfalfa serve as a good example of the close association between plant and pollinating organism. In this species, several mechanisms exist to ensure cross-pollination. The legume floral morphology is described in Chapter 1 and illustrated in Figure 1-11. The anthers normally burst before the flower buds open, while they are still green. This is a mechanism which is directed toward self-pollination, and in some species, including sweet peas and field peas, self-pollination does take place. In alfalfa and in the clovers, self-pollination is prevented by a cuticular membrane which covers the stigmatic surface. In white clover there is a substance in the style tissue which prevents the growth of the plant's own pollen. As the pigmentation of the flower develops, the stigmatic column is retained inside the two keel petals, the edges of which are usually held together by small hooks. The stigma presses upward against those parts of the keel petals which are below the standard petal [Figure 15-1(A)].

Bees which visit the flower to collect both nectar and pollen push aside the hooks which hold the keel petals together. The stigmatic column then springs upward, striking the bee and covering the lower part of its body with pollen. At the same time, the bristles on the bee's exoskeleton rupture the cuticular membrane covering the stigma [Figure 15-1(B)]. Pollen cells may then come in contact with the stigmatic surface and pollen germination will start.

The pollen is of two types, either from the same plant or from a plant previously

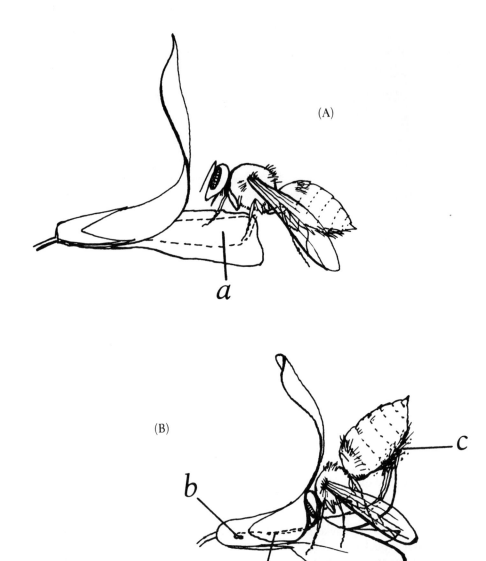

a. keel petals containing staminal column
b. nectary
c. stigmatic surface
d. keel petals
e. bee's proboscis

FIGURE 15-1
An alfalfa flower and a leaf-cutter bee. (A) Flower and bee before the flower is tripped. (B) Leaf-cutter bee tripping an alfalfa flower.

FIGURE 15-2
Leaf-cutter bee leaving the alfalfa inflorescence after tripping the two florets on the right. (Source: K. W. Richards, Research Station, Agriculture Canada, Lethbridge, Alberta)

visited by the bee. The plant's own pollen tube grows slowly, or not at all, down the hollow style toward the ovules. The growth rate of foreign pollen tubes is more rapid than that of the plant's own pollen and, hence, cross-fertilization usually takes place. When the bee leaves the flower, the stigma is carried forward and upward from the position shown in Figure 15-1(B). It presses against the standard or banner petal so that no further pollen can reach the stigmatic surface (Figure 15-2). Fertilization takes place 24 to 32 hr after the pollen germinates. The stigma remains receptive for up to 2 weeks if the flower is not visited by a bee.

This pollination process is usually called *tripping*. It is of vital importance to the seed producer, since without it there will be little seed. Consequently, if legumes are grown for seed in an area where there are few wild bees or other pollinating insects, seed producers place beehives in their fields. Where honey bees will trip the legume flowers, they are used, since they also provide a commercial by-product. In the case of alfalfa, honey bees are not attracted by the pollen and, in addition, are able to extract nectar without tripping more than 2% of the flowers. Also, in some areas wild bees are attracted to weed flowers or the flowers of other agricultural crops, rather than to alfalfa. Under these circumstances, leaf-cutter bees (*Megachile rotundata* Fabricius) are frequently used for cross-pollinating alfalfa.

(A)

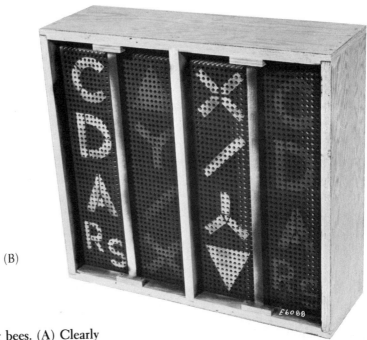
(B)

FIGURE 15-3
Nesting boxes for leaf-cutter bees. (A) Clearly marked hive visible to the bees from all parts of the field. (B) Distinctive markings on the nesting box enable the female bee to locate the nest tunnel on which she is working. (Source: K. W. Richards, Research Station, Agriculture Canada, Lethbridge, Alberta)

Leaf-cutter Bees

These small Mediterranean bees emerge from leaf capsules which have been formed from pieces of plant leaf. The leaf capsules may be purchased and stored at 4.5°C (40°F) until about 3 weeks before the bees are required in the field. Incubation at 30°C (86°F) leads to the emergence of the males, followed by the females. After mating, the male plays no further part in pollinating the alfalfa flower. The female, having no wax gland, takes pieces of leaf, makes them into capsules, and places them in the hive or nesting box provided [Figure 15-3(A) and (B)]. These should be distinctively marked so that the bees may locate them easily and should be spread throughout the fields, since the bees work close to the nest. The female fills the capsule with alfalfa pollen and some nectar, lays an egg, and closes the capsule (Figure 15-4). In one season, a female fills five nest tunnels, tripping alfalfa flowers as she collects the pollen. The *larva* in the capsule feeds on the pollen and nectar over the winter and, under natural conditions, emerges in the spring. Not all strains of leaf-cutter bees make satisfactory pollinating organisms in northern areas with cool summers, since they require temperatures of 20°C (68°F) or over before they can fly. In recent years, strains of leaf-cutter bees which will work at lower temperatures have been bred at the Agriculture Canada Research Station at Beaverlodge, in northern Alberta.

In the United States, a native wild bee, the alkali bee, has been found to trip alfalfa flowers successfully. The alkali bee is a ground-nesting species and has proved difficult to culture. However, in the intermountain states where it is naturally abundant, it is a valuable pollinating insect and plays an important role in alfalfa seed production.

FIGURE 15-4
Exposed nest tunnels showing capsules. (Source: K. W. Richards, Research Station, Agriculture Canada, Lethbridge, Alberta)

SEED TRADE ORGANIZATIONS

So far in this chapter we have seen that, especially for the grasses, it is important that seed production take place in areas with low populations of disease and insect pests. While appropriate seed treatment controls most pests and diseases borne externally on the seeds of legume crops, some diseases, mainly viruses, cannot be controlled in this way. Frequently, such diseases are spread by insect vectors. Cold climates, either at high altitudes or in northern regions, preclude the buildup of such insects. Additionally, when the climate is cool, the low temperature requirements for induction in the grasses are met. Thus, such regions are especially suitable for both grass and legume seed production.

As with other crops, seed production for forages calls for a knowledge of special techniques, many of which are set out in government legislation and controlled by means of government inspection. The need for such regulations arises from the diverse breeding systems (including pollination mechanisms) of the different plant species and the way in which plant breeders manipulate these systems in their efforts to improve the yield and quality characteristics of the different species. In the following pages, we will consider the plant-breeding practices which influence seed multiplication.

Plant-Breeding Practices

Most of our forages are cross-pollinated perennial crops which are grown in mixtures. They are expected to give high yields under environmental circumstances which are more diverse than for any other type of crop plant. The nutritional quality of their herbage and their resistance to diseases and insect pests are also of importance. Plant-breeding programs aim to produce improvements in these characteristics and to select plants in such a way that they express these improvements over the wide range of environmental conditions under which forages are usually grown. The potential new cultivars so produced are tested at many different locations and, in view of the perennial nature of the plants, over a number of years. Many forage-breeding programs, and the new cultivar-testing programs which are associated with them, are very complex. However, the most widely used forage-breeding method is that of *mass selection* applied to genetically diverse material which may have been obtained from an existing cultivar, from material collected in the field, or from crosses.

The smooth bromegrass breeding program at the University of Alberta will serve as an example of a typical forage-breeding program using mass selection. The initial source of genetic variation for the program was obtained by collecting clonal material from farmers' fields in Alberta. This approach was used because almost all the cultivated forage species grown in western Canada were introduced from either Europe or Asia. These plants have been grown in a new and extreme climate, with long, hard winters and dry summers, for 25 to 80 years. Both natural selection and adaptation take place in such materials. Consequently, the genetic variation present in old, established pastures (35 years old or more) was considered to be the most appropriate starting point for a plant-breeding program.

Clonal material was collected from fields in the summers of 1970 and 1971 (Figure

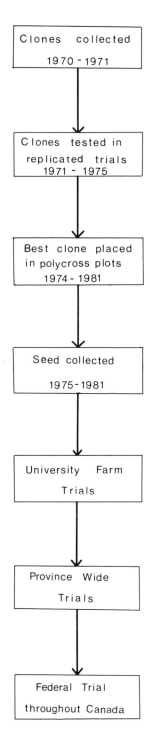

FIGURE 15-5
Stages in the smooth bromegrass breeding program at the University of Alberta.

15-5). Each of the plants was divided into several clonal pieces, which were established at the University Farm in replicated trials. There the clonal pieces were compared during the summers of 1971 to 1975 with similar material from established commercial cultivars. Plants giving high forage yields of good-quality herbage were selected and placed in *polycross* or *panmixis* breeding plots. Such plots are grown in isolation from all other plants of the same species and permit the interpollination of plants which have consistently given outstanding performances. In the University of Alberta breeding program, the first polycross plots were established in 1974. New plots containing various combinations of selections from the test plot were established yearly up to 1981. Some of these plots are still maintained.

The seed obtained from the polycross plots is regarded as the basic seed stock of a potential new cultivar which at this stage is called a new *synthetic*. This seed is used to establish cultivar tests in which the new synthetics are compared with existing commercial cultivars. Where the new synthetics give favorable forage yields, the tests are repeated at a number of locations in the Province of Alberta. If the synthetic strains are still successful, they are tested further in federal uniformity trials conducted throughout Canada. At this stage, seed production, which has been under review throughout the test period, will be considered along with yield and quality of herbage. On the basis of the information gathered from all these types of tests, an application is made to the Plant Products Division to license the new synthetic as a new cultivar (Figure 15-5).

In breeding programs of other types, the early testing and selection stages are more complex. Different crossing systems are used to test the genetic composition of the various clones before placing them in polycross plots. Top crossing, in which a number of clones are each crossed to a single clone, and diallel crossing, in which each of 7 to 14 different clones is crossed in all possible combinations with each other, are examples of systems which seek to investigate the plant's genotype. In mass selection programs, only the phenotype is studied.

The amount of seed produced from the panmixis or polycross plots is relatively small. The production from a single season will vary, depending on the environment and the productivity of the strain, and may amount to between 2 and 20 kg (about 4 to 40 lb). This seed is called the *first synthetic seed* or S_1. In the same way as the product of a cross known as F_1 (or first filial generation) gives rise subsequently to F_2, F_3, F_4, and F_5 generations, so the S_1 seed gives rise to S_2, S_3, S_4, and S_5 generations.

The parental clones used to develop these synthetic strains are composed of plants which are genetically diverse and heterozygous. For example, even if all the clones are high yielding, the genetic pathways by which they achieve this yield level may be many and varied. The cross-pollination which takes place between individual plants as the material moves from one generation to another results in the re-sorting and reallocating of the different genes. Groups of genes which interact (i.e., show epistasis) in the parental clones could be dissociated in the S_1 or subsequent generations. Most frequently, the outcome of this redistribution is a gradual decline in the character for which selection was made (e.g., yield) from S_1 to S_2, S_2 to S_3, and S_3 to S_4, continuing in like manner as the generations advance.

This genetic phenomenon has been called *genetic homeostasis*. The explanation of this change, in its most simple terms, is that, while it is possible to change the expression of a character in a population in a desired direction (e.g., to increase yield), there is a tendency for the trait to revert to its former level once the selection pressure which caused the change is relaxed. Applying selection pressure for characters such as yield is analogous to stretching out an elastic band which will spring back when the pressure is removed. The degree to which such reversions take place, if at all, will depend on the nature of the genetic control of the trait under selection.

The nature of the genetic control of yield, quality, and disease and pest resistance in forage species is such that changes in the expression of these characters frequently take place when selection pressure is relaxed. It is therefore most important that S_1 seed be multiplied in such a way that it is moved through as few S generations as possible before a large enough bulk of seeds is available to permit distribution of the new cultivar to the farmer. Federal testing of new cultivars is usually carried out using S_2 material, and the seed frequently reaches the farmer in the S_4 or S_5 stage.

Seed Multiplication

The objective of a seed-multiplication scheme is to take the small amount of seed which the plant breeder produces and increase this for distribution to farmers. These schemes are regulated by government legislation; the Canada Seeds Act makes provision for such control. The details of the way in which this legislation is to be implemented are set out in an appendix (Circular 6) to the Seeds Act. This is revised and updated from time to time. The act aims to achieve rapid seed multiplication, to avoid contamination (genetic or mechanical) with other cultivars or other crop species, to reduce weed seed contamination, and to keep disease and insect infestations to a minimum. It also makes provision for maintaining a pedigree of all seed stocks being multiplied in the scheme and for the inspection of such crops. In addition, it stipulates that certificates be issued to identify the material and supply information concerning seed purity. The organizations which are responsible for implementing the Canada Seeds Act are, first, the Plant Products Division of Agriculture Canada, which also maintains the pedigree records, and, second, the members of the Canadian Seed Growers Association, who grow the seed crops and organize the distribution to their members of seed available for multiplication.

In the United States, similar legislation exists to safeguard the seed purchaser. The Federal Seed Verification Service was established in 1927 and a Federal Seed Act became law in 1939. In 1940, the National Foundation Seed Project was initiated to help the foundation seed organizations through the state agricultural experiment stations, thus assuring adequate supplies of pedigreed seed.

An international seed certification plan was introduced in Canada, the United States, and Europe in 1950. Under this plan, the Organization for Economic Cooperation and Development (OECD) records the pedigree of seed moving in international trade and provides certificates which identify such material. To achieve this, the organization provides for field and seed inspection, and checks production, harvesting, and seed cleaning. The OECD follows rules and procedures which are very similar to those of national seed-certification agencies; this organization has helped to standardize terminology among different countries. For example, both Canada and the United States recognize four types of seed. For forage crops, these are, for native cultivars,

TABLE 15-2
Row spacing and seed rates for forages grown for seed production

	Row spacing		Seed rate	
	cm	ft	kg/ha	lb/acre
Alfalfa	40	1.2	4	3.5
Alsike clover	40	1.2	2	1.75
Crested wheatgrass	60–90	2–2.75	2–3	1.75–2.5
Red clover	30–45	1–1.5	2–4	1.75–3.5
Russian wildrye	90–120	2.75–3.5	3–4	2.5–3.5
Smooth bromegrass	60–90	2–2.75	2–3	1.75–2.5
Sainfoin	30–90	1–2.75	7–20	6.25–18

breeder seed (S_2), *foundation seed* (S_3), and *certified seed* (S_4). Cultivars introduced from outside the country are placed in the *registered seed* category.

In the United States, the enactment in 1970 of Public Law 91-577, Plant Variety Protection Act, provided a means of protecting new forage varieties. The breeder of a new variety is paid royalties in much the same way as is the author of a book and may control the new cultivar in the manner that a patent enables an inventor to control a product. This legislation encourages participation in cultivar breeding by private industry. Similar legislation (usually called "plant breeders' rights") is now in place or under consideration in many other countries.

Commercial Seed Production Practices

When forage plants are grown for seed production, they are normally planted in rows 40 to 120 cm (14.5 to 46.5 in.) apart, using a lower seed rate than is common for forage production (Table 15-2). In some cases, cross-blocking is practiced to remove part of the row. This procedure has been shown to be beneficial for alfalfa and grass-seed production. In stands produced by cross-blocking, weed control is easier than in solid stands and, with some species, seed is produced in the first or establishment year.

For grass seed production, nitrogen fertilization is most important. Grasses which develop mainly in the summer respond well to the spring application of nitrogen. Grasses that start their floral development in the winter (e.g., many fescues) require a split application, in the early fall and in spring.

Weed control is important in forage-seed production. Many difficult seed-cleaning problems may be avoided if the weed seeds are not present in the first place. The seed producer not only incurs the expense of the cleaning operation, but there is also a loss because some of the forage seed is usually removed at the same time as the weed seeds. Small weeds are destroyed by harrowing between the rows. In grass seed production, chemicals are also valuable in controlling weed infestations (see Chapter 14). The management of postharvest crop residues also influences seed yield by protecting the crop from pests (Table 15-3).

TABLE 15-3
Influence of postharvest residue management on the seed yield of three grass species in Washington State

	RESIDUE TREATMENT		
	Burned	Straw and Stubble removed	Straw removed
METRIC	Seed yield (kg/ha)		
Red fescue	636	539	459
Smooth bromegrass	1,122	914	848
Crested wheatgrass	872	716	790
ENGLISH	Seed yield (lb/acre)		
Red fescue	566	480	409
Smooth bromegrass	999	813	755
Crested wheatgrass	776	637	703

(Source: Canode and Law, 1978, Agron. J. 70. By permission of The American Socity of Agronomy, Inc.)

Harvesting forage seed presents a difficult problem. Frequently, all the seed on a plant does not ripen at the same time. The legumes, especially, have an indeterminate growth habit, and new flowers may be forming when seed from earlier flowers are mature. In the trefoils and in some grasses, this early seed will be shed before subsequent seed is fully formed. Harvesting is carried out either by direct combining or by combining from a windrow (see Figure 6-2). The use of foliar dessicants to terminate growth will facilitate direct combining.

Following harvest, the seed is dried, cleaned, and tested for germination. Cleaning involves the removal of stalks and pieces of stem, as well as weed seed, the seed of other crops, and foreign material. Seed storage conditions are of considerable importance. The higher the relative humidity and temperature, the more quickly the germination percentage will fall. Dry seed may be stored in moisture-proof containers.

FURTHER READING

Bohart, G. E. 1957, Pollination of alfalfa and red clover, Ann. Rev. Entomol. 2:1–28.

Bolton, J. L., 1962, Morphology and Seed Setting, Alfalfa, Interscience, New York, pp. 97–114.

Garrison, C. S. 1960, Technological advances in grass and legume seed production and testing, Advan. Agron. 12:41–128.

Rogler, G. A., H. H. Rampton, and M. D. Atkins, 1961, The Production of Grass Seeds, U.S.D.A. Yearbook Agriculture.

Rollins, S. E., and F. A. Johnston, 1961, Our laws that pertain to seeds. In Seed, U.S.D.A. Yearbook Agriculture, pp. 482–92.

chapter 16

Palatability and Grazing Behavior

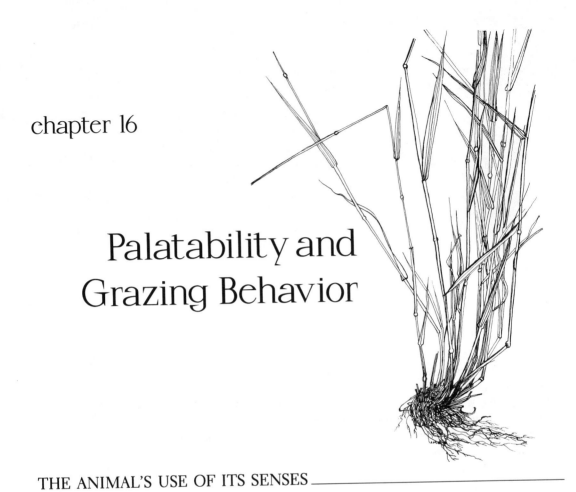

THE ANIMAL'S USE OF ITS SENSES

"Palatability" is a term which is widely used to express the preferences which livestock show for one forage over another. This usage is unfortunate, since it indicates that it is mainly taste which determines the animal's choice of foods. In fact, of the animal's five senses, at *least* three play a role in forage selection. Smell, touch, and taste all appear to be of equal importance. Sight may help to orient the animal in relation to the vegetation, but since all animals, except primates, lack color vision, its use in making food choices is limited. The importance of touch, taste and smell can easily be illustrated by examples of situations well known to anyone acquainted with farm animals.

Smell

Cattle will not eat the green, vigorously growing grass around dung pats. However, if the grass is cut and moved to another location, if the dung is sprayed with a chemical which will mask its smell, or if the animal's olfactory organs are anesthetized, such material is readily eaten. Obviously, smell is the sense determining preference in this case.

Touch

Livestock will not eat rough, harsh, or spiny material. The epidermis of a cow's muzzle has nerve structures like those found in human gums. When this area, together with the mouth parts, is anesthetized, cattle will eat coarse materials (including thistles) which are normally rejected. Under these circumstances, touch is the sense which influences selection.

Taste

There is considerable evidence that taste is a sense extensively used by livestock for food selection. Both cattle and sheep will relish material previously unacceptable if it is sprayed with molasses or saccharin. It has been shown that even goats, well known for their lack of discrimination between food types, can distinguish the four primary tastes (sweet, bitter, sour, and salt).

In general, we may conclude that farm animals are *linked* to their food substances by both chemical and physical stimuli received by their senses. The chemical stimuli with which their senses are bombarded are many and varied. The animal's response to this vast number of chemical stimuli it receives when grazing on pasture may be either behavioral or physiological. However, it is always selective. This selection of *significant* stimuli will differ from one animal species to another and also among individual animals. Thus, the degree of a response is determined not only by the intensity of the stimulus (e.g., for a chemical stimulus, the number of molecules reaching the animal), but also by the number and nature of the receptors. It is these structures that vary in number and type from species to species and among animals within a species. Furthermore, work by Arnold (1966) indicates that food preferences are rarely the outcome of a response to only one sense. Receptors of different types must reinforce the information which they provide for the animal before it will show a marked food preference. Such reinforcement is frequently lacking so that animal food preferences are poorly expressed, thus making selection and rejection thresholds difficult to determine.

Forage preference or *palatability* has been defined as the sum of the factors which determines the degree to which food is attractive to an animal. Thus, animal factors such as age, health, condition, and environment form a part of "palatability," along with the physical and chemical nature of the food substances.

EVIDENCE FROM EXPERIMENTS AND EXPERIENCE

Since animal food preferences are difficult to determine, and since the theoretical background which might aid our understanding is incomplete or lacking entirely, we need to be aware of the evidence now available to us from both experimentation and experience.

TABLE 16-1
Percentage composition of esophageal fistulae samples collected from six steers

Species	\multicolumn{6}{c	}{Steer No.}	Mean
	1	2	3	4	5	6	
Red top	—	25.3	34.5	19.4	—	9.9	14.9
Crested wheatgrass	21.7	3.2	2.6	2.1	—	1.3	5.1
Intermediate wheat	11.1	2.5	1.6	16.0	7.0	4.8	7.2
Bromegrass	39.6	33.3	38.7	19.1	9.7	17.0	26.2
Russian wild ryegrass	—	—	—	—	1.2	1.9	0.5
Creeping red fescue	15.6	29.7	7.5	19.3	27.5	42.6	23.7
Birdsfoot trefoil	1.4	4.2	1.3	1.3	17.2	0.7	4.4
Alfalfa	10.6	0.3	4.2	15.8	—	18.1	8.2
Sainfoin	—	—	—	7.0	—	0.8	1.3
White clover	—	1.5	9.6	—	37.4	2.9	8.6
Least significant difference: 5%							11.82
1%							15.82

(Source: R. H. Gesshe, 1978, M. Sc. thesis)

The Individual Animal and the Herd

At the University of Alberta ranch, a small herd of animals, which consisted of six steers with esophageal fistulae, were put to graze in a pasture which contained replicated half-acre plots of the ten forage species listed in Table 16-1. The mean percentage composition for the material collected from the fistulae indicates that the herd as a whole preferred smooth bromegrass and creeping red fescue. There are, however, individual animals who ate very little of either of these forages. For example, the diet for steer No. 5 contained only 9.7% bromegrass, while steer No. 3 ate only 7.5% creeping red fescue. In contrast, redtop formed 34.5% of the diet of steer No.

TABLE 16-2
Steer preference rating for grasses and legumes grown on the University of Alberta Ranch at Kinsella

Stage of development	Grasses	Legumes
Vegetative (May 31–June 15)	1.06	0.78
Heading and flowering (June 21–July 10)	0.60	1.04
Seed set (June 11–July 31)	0.73	1.40

(Source: R. H. Gesshe, 1978, M.Sc. thesis)

3, while steers No. 1 and 5 rejected this species entirely. Thus, there are marked preference differences between individual members of the same species. It is not surprising that, similarly, there are differences between species. For example, white clover, which is well liked by sheep, may be rejected by cattle.

Plant Species and Their Growth Stage

The stage of development of a pasture influences the animal's preference for the forage species of which it is composed. On the University of Alberta ranch at Kinsella, the preferences of steers grazing on the ten forage species listed in Table 16-1 were different at the vegetative stage (May 31 to June 15), at the time of heading and flowering (June 21 to July 10), and at the time of seed set (July 11 to July 31). Such preferences may be quantified by using the system of preference rating devised by Van Dyne and Heady (1965) and modified by Rosier et al. (1975). This preference rating is calculated for each individual species by expressing utilization and production as a percentage of the overall utilization and production for the pasture area as a whole. The percentage values for utilization are then divided by those for production. Ratings so generated with values *greater* than 1.0 indicate *preference*, while values *smaller* than 1.0 show *avoidance*.

Preference ratings of this type (see Table 16-2) show that, when grazed in the vegetative stage, the grasses listed in Table 16-1 are preferred over the legumes. In the second grazing period (June 21 to July 10), this situation is reversed. Considering next the individual forage species, during the first grazing period, the animals reject white clover, birdsfoot trefoil, and sainfoin, and prefer bromegrass, intermediate wheatgrass, and Russian wildrye. In contrast, Russian wildrye, together with birdsfoot trefoil and alfalfa, are preferred species during the second grazing period, while crested wheatgrass, intermediate wheatgrass, and sainfoin are avoided. During the third grazing period, when the plants are setting seed, creeping red fescue, crested wheatgrass, intermediate wheatgrass, and sainfoin are rejected entirely.

Choice of Plant Parts

Grazing animals select not only between forage species, but also choose between parts of the same plant. Upper leaves are preferred to leaf-bearing stems, and these, in turn, are selected over main stems. In grasses, some heads in the earlier stages of development are attractive to cattle (e.g., Russian wildrye), while the heads of awned species (e.g., rye) are rejected. Smooth bromegrass heads are rejected when mature. The amount of forage available and its ease of access will also influence animal preference. If forage is limited, previous preferences will be abandoned, and material which is coarse, unpalatable, or even poisonous will be eaten.

Influence of Previous Diet

Previous exposure to a food substance may render it either more, *or less*, acceptable to livestock. For example, cattle will reject silage initially, but after it has been offered for 1 or 2 weeks, they will eat it readily. It seems that the rejection threshold concentration for silage in the diet will increase, possibly as a result of substances accumulating in the animal's body or change in rumen microflora. The animals are said to develop a "taste" for silage.

On the other hand, a herd of dairy cattle which showed a marked preference for an orchardgrass–white clover mixture over ten other simple grass–legume mixtures rejected this pasture in favor of a meadow fescue–alfalfa mix when they returned to the test site following one month spent grazing only orchardgrass and white clover. In this case, previous exposure resulted in rejection. The animals preferred "a change."

Management

A number of management factors, foremost of which is grazing intensity, have a profound effect on animal acceptance of forages. High grazing intensities will eventually increase the hunger of the animals so that rejection thresholds for both taste and smell are lowered, and materials which were unacceptable under conditions of plenty are eaten. Also, varying amounts of less palatable species may, under some circumstances, be eaten where associated with a palatable species. Under conditions of nitrogen deficiency, the application of a nitrogenous fertilizer will increase palatability.

A management practice which presents some dangers is the use of herbicides to "spot" treat poisonous plants in a pasture. Some cattle instinctively avoid poisonous plants, responding to the plant's smell. The use of a herbicide will mask the smell and block this warning mechanism. For this reason, cattle should be kept out of a pasture in which herbicides have been used until the treated weeds have died.

Quality Characteristics of Forage

There is evidence from both native (Hilton and Bailey, 1972) and cultivated plant species (Gesshe and Walton, 1981) that cattle will select plants with a high moisture content during the drier times of the growing season. Under these circumstances, grasses showing drought resistance will be overgrazed. Studies at the University of Alberta ranch show that, in general, animals will select material high in crude protein, low in crude fiber, and with high digestibility (Gesshe and Walton, 1981). Such evidence is supported by the results obtained by many workers who have found that material from esophageal fistulae have higher crude protein values than those obtained from samples clipped from the pastures in which the animals grazed. This should not be taken to indicate that livestock have some kind of "nutritional wisdom." A more likely explanation is that the feedback mechanisms that control selection and rejection thresholds are poorly understood.

GRAZING BEHAVIOR

The grazing habits of cattle take the form of a series of cycles, which may be divided into activities of three kinds: grazing, ruminating, and idling. The animals start at sunrise, or a little before, with their longest period of grazing (about 2 hr) and a very brief period of idling, followed by ruminating and then a longer period of idling. The total time spent on this first cycle is about 4 hr. Four or five shorter cycles (grazing–ruminating–idling) follow through the remainder of the daylight hours, with at least one grazing cycle after sunset.

For most breeds of cattle, the total time spent grazing is about 8 hr/day, but there is some variation in the daily duration of grazing that may be attributed to breed differences. For example, Aberdeen Angus and Hereford cattle graze for about 8.1 hr/day, while the Fulani cattle of northern Nigeria graze for only 6.7 hr. Table 17-8 shows the marked difference between the time which animals spend grazing on a poor pasture (10 hr/day) compared with the time they spend on a good pasture (8 hr/day). The animal's hunger will not drive it to graze beyond 10 hr/day, an increase of 25% over the normal 8 hr of grazing per day. During the normal 8-hr grazing time, the animal spends 5 hr eating, 3 hr "looking" for food, and walks about 2.5 miles. Where the total grazing time is longer than 8 hr, as on poor pasture, it is the searching or "looking" time which is increased (up to about 5 hr), while the time spent eating remains unchanged. In association with the average 8-hr grazing day, cattle spend 7 hr ruminating. Fiber content of the forage is positively associated with ruminating time. The remainder of the animal's day is spent resting, with about 12 hr of the ruminating and idling time being spent lying down.

The mechanical task that is presented to cattle in biting off (harvesting) their daily requirements of green material (70 to 95 kg; 154 to 209 lb) is formidable. Where pasture conditions are optimum (a dense stand of young grass, 10 to 13 cm or 3.5 to 4.5 in. high), the animal must take about 80 bites per minute to harvest 90 kg (198 lb) of green material in an 8-hr grazing day. Where the number of bites falls to 40 per minute, the intake will be 20 to 25 kg (44 to 55 lb) per day, which is scarcely enough for maintenance.

This vast task of harvesting must be carried out using the lower teeth, which clamp the grass against a muscular pad in the upper jaw, while a movement of the head shears the green material. This complicated harvesting task has to be carried out by teeth which have a total width of 6 to 6.5 cm (2.25 to 2.33 in.). To assist the animal with these substantial efforts, the height of the pasture is critical. If the pasture plants are too tall (i.e., 25 to 35 cm or 9 to 12.5 in.), the animal, by grazing from the lower parts of the plant, may take material about 30 cm long (about 11 in.) into its mouth. Forage of this size must be manipulated in the animal's mouth, thereby adding to the work of grazing, as well as reducing intake and weight gains.

HERD BEHAVIOR

The individual members of a herd tend to graze, ruminate, and rest simultaneously. If the herd is grazing, one or two animals who start to ruminate will not continue to do so unless the rest of the herd joins them. It would seem that one or two animals become herd leaders. Just what determines the "leadership" character is difficult to say. Where twin animals are separated to form two genetically identical herds, and the two herds are placed in identical pastures, they do not graze, ruminate, or idle at the same time. Evidently, different members of the twin pairs have become herd leaders for the two herds. This is taken to indicate that "leadership" is not an animal characteristic which is under genetic control and that some environmental influence is responsible. The behavioral sciences are relatively new, and there is much that is not known about the behavior of our domestic animals.

FURTHER READING

Arnold, G. W., 1966, The special senses in grazing animals, Aust. J. Agric. Res. **17:** 531–542.

Baile, C. A., and F. H. Martin, 1972, Effects of local anesthetics on taste and feed intake, J. Dairy Sci. **55:**1461–1463.

Bath, D.L., W. C. Weir, and D. T. Torell, 1956, The use of the esophageal fistula for the determination of consumption and digestibility of pasture forage by sheep, J. Animal Sci. **15:**1166–1171.

Conbett, J. L., 1966, The nutritional value of grassland herbage, International Encyclopedia of Food and Nutrition, Pergamon Press, Oxford.

Gesshe, R. H., and P. D. Walton, 1981, Grazing animal preferences for cultivated forages in Western Canada. J. Range Manage. **34(1):**42–45.

Hilton, J. E., and A. W. Bailey, 1972, Cattle use of a sprayed aspen parkland range, J. Range Manage. **25:**257–260.

Johnstone-Wallace, D. B., and K. Kennedy, 1944, Grazing management practices and their relationship to the behavior and grazing habits of cattle, J. Agric. Sci. **34:** 190–197.

Rosier, R. E., R. F. Beck, and J. D. Wallace, 1975, Cattle diets on semi-desert grassland: botanical composition, J. Range Manage. **28:**89–93.

Van Dyne, G. M., and H. F. Heady, 1965, Botanical composition of cattle and sheep diets on mature annual range, Hilgardia **36:**465–490.

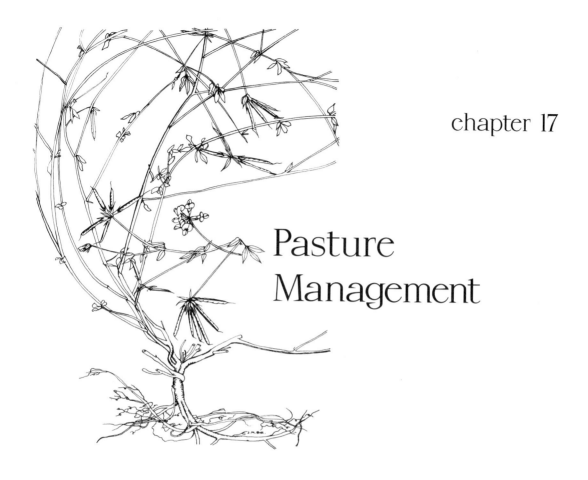

chapter 17

Pasture Management

ECONOMIC IMPORTANCE OF PASTURES

The economics of forage production have been widely studied to determine the cheapest way of maintaining livestock. While details vary, it is universally accepted that the use of pasture in grazing systems provides the most economical means of producing the quantity and quality of feeds required for livestock maintenance and production (Table 17-1). This is an important consideration for two reasons. First, the price of feed grain is high, and in many parts of the world the ruminant animal is regarded as competing with the human population for food. Second, ruminant animals are very inefficient in their conversion of feedstuff. For example, a steer gaining about 0.5 kg (1.1 lb) per day will eat more than ten times that amount of dry matter, giving a conversion efficiency of under 9% when estimated on a liveweight gain basis. However, if the carcass weight is used in the calculation, the conversion efficiency is less than 5%. It is only possible to operate an inefficient process of this type at a profit if the materials used are cheap. Also, since animal feed represents half the cost of meat production and three-quarters of the cost of milk production, it is essential that feed costs be minimized wherever possible. These considerations determine the importance of the role played by the pasture.

TABLE 17-1
Economic efficiency of forage production

Production system used to maintain livestock	Percentage cost		
	Netherlands	Canada	United States
Grazed pasture	100	100	100
Hay (all kinds)	140	152	160
Alfalfa hay	138	139	152
Timothy hay	143	167	161
Silage	187	193	195
Dehy (dehydrated forages)	294	281	320
Grain and concentrates	314	457	425

(Source: Agriculture Canada and U.S. Department of Agriculture publications. By permission of The American Society of Agronomy, Inc.)

ART OF MANAGEMENT

Pasture management is an art as well as a science. It calls for skilled and experienced judgments which cannot be learned either in the classroom or from a book. Such skills are developed by evaluating field situations. The objective is to obtain maximum livestock production while maintaining desirable species in a long-lived stand. The decisions that the pasture manager makes are, in fact, a series of compromises. First, he must balance pasture production against animal needs. Next, he must ensure that animal needs are met without detriment to pasture survival. Finally, he must determine stocking rates in such a way as to achieve optimum animal-weight gains per hectare, as well as the most favorable gains per individual animal. All this must be achieved against a background of considerable forage yield variation from year to year.

FORAGE YIELD VARIATION

On the University of Alberta Ranch at Kinsella, the average yield of dry matter from a pasture is 3,250 kg/ha (2,892 lb/acre) per year. However, this average is calculated from data that range from 5,010 kg/ha (4,459 lb/acre) in the best year to 1,570 kg/ha (1,397 lb/acre) in the worst. The factors that give rise to this wide range are well-illustrated by the data presented by Cook, Beacom, and Dawley (1965), part of which is shown here in Figure 17-1. Rainfall, the age of the pasture, and the time of fertilizer application have influenced, individually and in combination, production of the two types of pasture the authors consider. In addition to rainfall, other climatic factors greatly influence annual forage production. In the humid southern areas of the United States, average yields may well be as high as maximum yields in Canada, for, while soil fertility is frequently low in these areas, both rainfall and temperatures are favorable for much of the season. Consequently, where a nitrogenous fertilizer is added or a legume is present, forage production is high (Table 17-2). Figure 17-2 contrasts the carbohydrate reserves of pasture plants grown in two types of seasons. Cool nights, cool, cloudy days, and good soil moisture all tend to promote vegetative growth and delay flowering (year B). Warm nights, warm, sunny days, and low soil moisture levels have the reverse effect. Both carbohydrate reserves and forage yields are reduced when such conditions prevail (year A).

As well as between-season variation, within-season differences are very important. The major part of the herbage production frequently occurs in the early part of the season, as may be seen in data from Lethbridge, Alberta (Figure 17-3). However, feed requirements for steers increase, while those for dairy cows decrease slightly, during the grazing season. The dry-matter requirements for a range of livestock classes are given in Table 17-3.

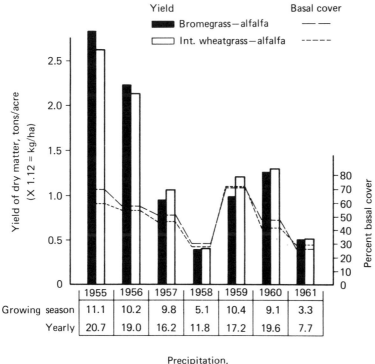

FIGURE 17-1
Dry-matter production and basal cover in relation to annual precipitation. (Source: D. A. Cook, S. E. Beacon, and W. K. Dawden, Pasture productivity of two grass–alfalfa mixtures in northeastern Saskatchewan, 1965, Can. J. Plant Sci. (45:167–168).

TABLE 17-2
Yields of dry matter for tropical grasses in the southeastern part of the United States

Nitrogen application	Bermudagrass	Dallisgrass	Bahiagrass
METRIC			
kg/ha of N		kg/ha	
0	2,713	3,274	2,425
120	5,897	4,175	3,719
180	6,974	5,732	4,721
With white clover	6,734	4,721	4,597
ENGLISH			
lb/acre of N		lb/acre	
0	2,414	2,914	2,158
107	5,248	3,716	3,310
160	6,207	5,101	4,202
With white clover	5,993	4,201	4,091

(Source: U.S. Department of Agriculture publication)

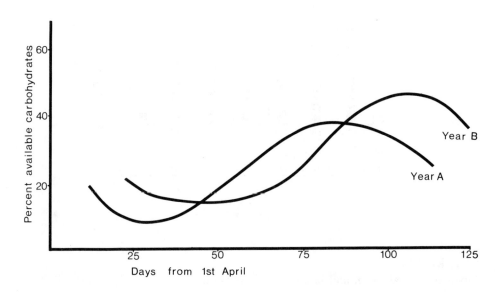

FIGURE 17-2
Storage and utilization of available carbohydrates in alfalfa roots.

TABLE 17-3
Herbage dry-matter requirements for a 120-day grazing season

Livestock	Dry-matter requirement	
	kg/120 day	lb/120 days
273 kg (600 lb) steer gaining 1 kg (2.2 lb)/day	1,086	2,389
273 kg (600 lb) steer gaining 1.5 kg (3.3 lb)/day	1,409	3,100
455 kg (1,000 lb) beef cow producing 18–23 kg (40–50 lb) of milk daily	1,441	3,170
64 kg (141 lb) ewe with 0.6 kg (1.3 lb) nursing lamb	1,966	4,325
23 kg (50 lb) lamb gaining 0.2 kg (0.5 lb)/day	360	792

(Source: Collected from various Agriculture Canada and U.S. Department of Agriculture publications)

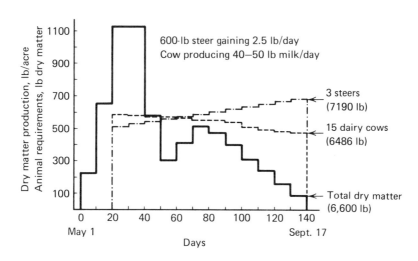

FIGURE 17-3
Herbage dry-matter production per acre and livestock requirements averaged over 10-day intervals. Conversion factors (lb × 0.45 = kg).
(Source: Canadian Forage Crops Symposium, 1969)

STOCKING RATE

Balancing available herbage supply with animal needs calls for the determination of stocking rates. This requires great skill. One way to make such an estimate is to compare the different classes of livestock with an *animal unit*, a standard equivalent to a mature (460 kg or 1,012 lb) cow which eats about 12 kg (26.4 lb) of forage dry matter per day (360 kg or 792 lb per animal unit month). The standard ratios in common use are given in Table 17-4. Stocking rates are frequently described as being "heavy" or "light." These terms are relative to pasture production. Light grazing indicates a stocking rate which will permit the accumulation of a substantial amount (usually 30% to 75%) of the top growth by the forage species. Under these circumstances, the animals are able to graze selectively, eating palatable species more heavily than less palatable ones. Heavy grazing involves stocking rates which will result in most of the herbage being removed. This is not necessarily overgrazing. There are some very simple ways in which herbage supplies may be balanced with livestock numbers after stocking rates have been determined: either surplus herbage, at the time of maximum growth, may be conserved for subsequent use, or supplementary grain may be offered in times of scarcity. The choice between these two possibilities is predetermined by the stocking rate in the early part of the season. If the pasture is underutilized at that time, with a stocking rate of the type shown in Figure 17-3, then part of the field may be fenced off and used for hay or silage. If, however, the early-season stocking rate is such that all the production is utilized, then supplementary feed is required later in the season. Both of these simple solutions to within-season variation in forage production introduce a more expensive feed system than that of using the pasture (Table 17-1). It is possible to avoid these high costs by managing the pasture in such a way that the plants are maintained in a juvenile or vegetative stage for as long as possible. The skill of a pasture manager may be determined by the degree of success with which this objective is achieved.

TABLE 17-4
Animal unit equivalents commonly used to determine stocking rates

Animal type	Animal unit equivalent
1 Cow (with or without calf at foot)	1
1 Yearling (steer or heifer)	0.65–0.75
1 Bull (mature)	1.5
5 Ewes (with or without lambs)	1
5 Does (with or without kids)	1
1 Mature horse	1.5

(Source: Collected from various publications)

PLANT GROWTH

The physiological factors which determine plant growth were discussed in Chapters 7 and 8. To recapitulate briefly: defoliation, either by cutting or grazing, not only removes organs that manufacture plant food (the leaves), but also reduces the activity and size of the root system. In perennial plants, either roots or horizontal stems (in legumes) or stem bases (in grasses) act as storage organs for reserve carbohydrate material. The vigor of the plant is thus reduced by defoliation, since the regions in which the food reserves are both produced and stored are adversely influenced. If the plant is defoliated too early in any season, too frequently, or too close to the ground, the reduction in vigor is such that the plant may not survive. In other words, heavy stocking, which leads to frequent, close defoliation, reduces carbohydrate reserves and leads to the death of those plants which the animals find most palatable and, hence, graze most frequently. Weeds and unproductive plants will then increase, thereby decreasing pasture quality and productivity.

PLANT CHARACTERS

Some plant species are able to withstand the detrimental effects of frequent and close grazing. These plants are either low in stature (e.g., Kentucky bluegrass) or have a substantial number of basal leaves (e.g., crested wheatgrass). For Kentucky bluegrass, 25% of the plant's height is less than 1 in. above the ground and is inaccessible to the bite of the grazing animal. In the case of crested wheatgrass, only 11% of the plant's total height is within 50 mm (2 in.) of ground level, but this region also contains 30% of the plant by weight. Heavily grazed areas, such as holding areas or corrals, consist of Kentucky blue grass and white clover in the more humid parts of the Canadian prairies. In the humid southern United States, summer perennial pastures can withstand continuous grazing during the early part of the season when the growth is lush. At that time, stocking rates may be double those used later in the season. From mid-June onward, growth slows and cattle numbers should be reduced. Rotational grazing (to be described in this chapter) is then practiced for the remainder of the season.

UNDERGRAZING AND OVERGRAZING

Undergrazing can be just as detrimental as overgrazing, but for different reasons. Where animals have more pasture than needed, they will move over the area available, selecting those plants and plant parts which they find most attractive. The ungrazed plants become mature, fibrous, and even less attractive and, hence, become less likely to be grazed. These ungrazed plants are better able to compete with the grazed plants for light, water, and soil nutrients. Consequently, the least desirable plants tend to spread and increase in the pasture. It is, therefore, most important that the pasture manager should achieve an optimum intermediate between under- and overgrazing. Such a balance is closely associated with the balance the pasture manager seeks between gain per animal and gain per hectare. Since production costs must be considered in relation to land values, the most realistic evaluation of cost for animal products is on a land area basis. However, many farmers consider production in terms of animal gains or output. The maximum gains per animal are obtained by understocking, which will be accompanied by pasture deterioration. There is, however, an intermediate optimum range where both gains per animal and per hectare are high, and which represents a satisfactory balance between under- and overstocking. Within this range, individual animal gains are high and pasture condition is preserved. This is important, for while the animal is of prime consideration, the pasture, too, is of value and, if allowed to deteriorate, is costly to replace.

REVIEW

To this point, we have shown that, first, forage production from pastures is highly variable both within and between seasons; second, it is important that pastures be neither under- nor overstocked if animal production is to be high and pasture condition is to be maintained; and, third, the most productive and nutritious pasture is one consisting of plants in the vegetative stage (see Chapter 9). To accommodate all these factors, a management system is required which has two characteristics: (1) it must be flexible, and (2) it should allow the pasture plants a rest period in which to recover from grazing.

GRAZING SYSTEMS

Many grazing systems exist which have both of the above-mentioned characteristics. These include seasonal, alternate, deferred, complementary, and rotational grazing. All these grazing systems may be applied equally well to the three main pasture types. These consist of, first, pastures which are part of an arable crop rotation, and consequently are plowed up from time to time (rotation pastures), second, permanent pastures, and, third, supplementary (or temporary) pastures, which often consist of annual forages such as a spring or winter grain crop.

Seasonal grazing is a system which is simple in concept, but sometimes difficult to achieve in practice. The different annual and perennial forages reach their peak production at different times of the year. Consequently, if a number of pastures can be established using appropriate species with production peaks following each other in sequence, high stocking rates may be maintained over the whole grazing season. A system of this type has been devised for southern Alberta (Figure 17-4). It can also be simplified and operates well with only three types of pasture (crested wheatgrass, native range, and Russian wildrye).

Alternate grazing, as the name suggests, employs two pastures, one of which is rested while the other is in use. Such an alternating system may operate using a summer

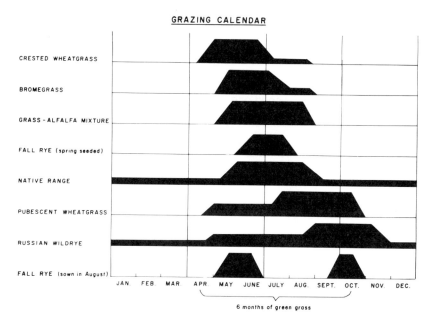

FIGURE 17-4
Peak production for grass species in southern Alberta. (Source: Agriculture Canada publications)

and winter pasture, which might consist of a cultivated pasture (bromegrass–alfalfa) in summer and native range in the fall, or fall and winter. Equally well, the alternation might be yearly between two pastures containing the same species. In the humid, southeastern part of the United States or in the drier areas of northern Mexico, it is possible to alternate between temperate grasses (tall fescue, orchardgrass, smooth bromegrass, reed canarygrass, or annual ryegrass), which give maximum production in the winter, and tropical grasses (bermudagrass, bahiagrass, dallisgrass, or johnsongrass), which give maximum production in the summer. For example, in the lower south of the United States, the annual species ryegrass, crimson or arrowleaf clover, field peas, and vetch are frequently used for winter production (November, December, January, February, March, April, and May) in conjunction with summer feed from perennials like kudzu, sericea lespedeza, and white clover, grown with bermuda grass, bahiagrass, or dallisgrass. The summer annual species frequently used are cowpeas, velvetbeans, and alyce clover *(Alysicarpus vaginalis)*. A tall fescue and white clover mixture is very productive in April, May, early June, and November. From this wide range of species, it is possible to develop many systems to provide feed all the year round. In Virginia, similar results might be obtained by growing Kentucky bluegrass and white clover, which will give two production peaks (in May and in September to October). Between these times, sudangrass hybrids could be used from June to September and for conservation for the winter months.

	Year 1	Year 2	Year 3
Grazed for first one-third of grazing season, then rested	A	B	C
Grazed for second one-third of grazing season, then rested	B	C	A
Grazed for one-third of grazing season, then rested	C	A	B

FIGURE 17-5
Deferred rotational grazing system for three pastures (A, B, and C).

The more elaborate systems developed from this method are frequently called deferred grazing or deferred rotational grazing systems (Figure 17-5). Systems of this type have been widely applied on the North American subcontinent in northern Mexico, in the southwestern United States, and in central and western Canada. Marked yield increases of up to eight times as much forage have been obtained using this method under semiarid grazing conditions.

Complementary grazing employs pastures of contrasting types. Extensive tracts of poorly producing land, often in hill areas, are used in conjunction with smaller, highly productive areas which are usually fertilized and often irrigated. Such systems have been used in western Canada since 1935 and are common in the hilly areas of Scotland and England, in the United States, and in Mexico.

In many ways, all these grazing systems may be regarded as extreme variants of rotational grazing, a particularly flexible system which has been adapted to environments ranging from northern Canada to the tropical areas of South America. Rotational grazing has also been successfully used in Europe, Asia, and Africa.

ROTATIONAL GRAZING

The principles of this system are best illustrated by an example. Assume that a field of about 16 ha (40 acres) is to be rotationally grazed with 40 steers, averaging about 275 kg (605 lb) each. The field could be divided into four equal areas with all the animals being placed in one of these 4 ha (10 acre) fields. When the forage available in this area has been grazed off, the animals are moved to the next 4 ha (10 acre) pasture. This process continues until all four fields have been grazed.

By doing this, we have confined a large number of animals to a relatively small area for a short time. Under these circumstances, the whole area is grazed quickly and uniformly. There is less waste by trampling and fouling and no opportunity for selective grazing. The period of occupation of a pasture may be varied to accommodate the forage available in the pasture, the animals being moved every 5, 10, 14, or 30 days, depending on prevailing conditions. The animals are moved to a new pasture either when the grass has been eaten down to a predetermined level or when vegetation in the next pasture has grown to a suitable height for grazing. In either case, since the growth rate of the grass will vary over the growing season, so will the duration of occupation of any single unit.

The most important feature of rotational grazing from a pasture point of view is the duration of the period when the pasture is *not* occupied by the grazing animals. This is the time when the plants, which have been uniformly grazed, can build up their photosynthetic area again, when carbohydrate reserves used in the early stages of regrowth are replenished, and when root systems recover from the effect of defoliation. It is important that this rest period be long enough to produce young vegetative growth which can replenish carbohydrate reserves, but be short enough to prevent the grass from becoming coarse or starting to mature and head.

In the early part of the grazing season, when production is high and growth rates are rapid (see Figure 17-3), the flexibility which may be provided by lengthening the period of occupation of any one pasture may result in other pastures becoming too mature. Under these circumstances, it is possible to remove one of the fields from the grazing sequence and harvest it for hay or silage. Such a field would be introduced into subsequent rotational cycles and would provide feed for livestock later in the season at a time when growth rates had slowed.

Example of Rotational Grazing

The duration of occupation and numbers of rotational cycles have been determined for the University of Alberta Ranch at Kinsella (Table 17-5). The animals are placed in the first field in mid-May, when the herbage might be adequate for only 3 days of grazing. As the growth rates increase rapidly to a maximum, the animals remain

TABLE 17-5
Times of occupation and rotational cycles for a four-field rotational grazing system

Days per cycle	Total grazing days	Fields used and duration of occupation	Rotation and cycle
17	17	1 field, 3 days 2 fields, 7 days	1st
21	38	3 fields, 7 days	2nd
32	70	4 fields, 8 days	3rd
40	110	4 fields, 10 days	4th

in each field for a longer period (about 7 days). In the initial rotational cycle, only three of the four available fields are used. During this time, the condition of the pasture to which the animals are next to be moved determines the time when the move takes place. Under these circumstances, a reasonable amount of herbage (about 10 cm or 3.5 in. of growth) would remain in the pasture after grazing had finished. As a result, regrowth starts more rapidly (see Chapters 7 and 8), and a substantial supply of forage is available for subsequent rotational cycles. This additional material results in an improved growth rate at a time when the growth rate would otherwise be slow (see Figure 17-3). The field which was used for silage in the second grazing cycle (see Table 17-5) is introduced into the grazing rotation in the third cycle. During the third and fourth cycles, the condition of the pasture occupied by the animals determines the time when the animals are to be moved.

This type of management system changes the pasture production pattern in such a way that forage is available to maintain continuous liveweight gains throughout the grazing season. This is illustrated by data from a trial carried out at the University of Alberta Ranch at Kinsella between 1975 and 1978, which compared continuous stocking with rotational grazing of the type considered here. Figure 17-6 shows that, while

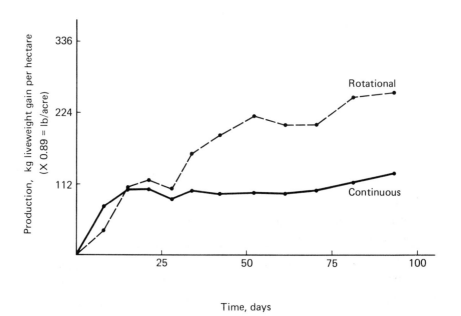

FIGURE 17-6
Weight gains from continuous and rotational grazing, University of Alberta Ranch, 1977. (Source: P. D. Walton et al., 1981, J. Range Manage. 34:19–21)

weight gains under continuous grazing followed the forage production pattern shown in Figure 17-3, animal gains from rotational grazing were consistent throughout the grazing season. By the end of the season, these gains were double those obtained from continuous grazing. The difference between the animal weight gains from the two grazing systems was mainly due to the differences in the amounts of forage consumed (see Tables 17-6 and 17-7). The forage conversion ratio was relatively consistent for the two grazing systems. Forage production was high under rotational grazing, supporting high stocking rates and giving high daily gains. The influence of rotational grazing on livestock performance was most marked after the second year.

Grazing management also influences animal behavior. While there were no significant differences between the times which the animals spent ruminating under the two grazing systems (Table 17-8), the hours spent grazing were substantially greater under continuous than under rotational grazing. This difference increased as the season progressed; the animals under continuous grazing spent 0.8, 1.5, and 1.8 hours longer grazing in June, July, and August, respectively, than did the rotationally grazed animals. A linear relationship between increased grazing time and decreased forage availability has been reported for both sheep (Arnold, 1960) and cattle (Lofgreen, Meyer, and Hull, 1957). The increased grazing time shown for continuous grazing no doubt accounts in

TABLE 17-6
Forage consumed and weight gains under continuous and rotational grazing, Kinsella

METRIC

	Forage consumed kg/ha		Liveweight gain kg/ha	
	Continuous	*Rotational*	*Continuous*	*Rotational*
1975	2,554	2,703	266	318
1976	2,291	2,173	166	275
1977	1,012	2,045[a]	119	235[b]
1978	1,015	2,081[a]	118	202[c]

ENGLISH

	Forage consumed lb/acre		Liveweight gain lb/acre	
	Continuous	*Rotational*	*Continuous*	*Rotational*
1975	2,273	2,406	237	283
1976	2,039	1,934	148	245
1977	901	1,820[a]	105	209[b]
1978	903	1,852[a]	106	180[c]

[a]Unpaired t test (d.f. = 4) shows significant differences between continuous and rotational grazing at the 5% level of probability.
[b]Significant at 1% level of probability.
[c]Significant at 0.1% level of probability.

(Source: P. D. Walton et al, 1981, J. Range Manage. **34**(1):19–21)

TABLE 17-7
Conversion ratio, average daily gain per animal, and stocking rate under continuous and rotational grazing

METRIC

	Conversion ratio kg of forage/kg of beef		Average daily gain kg		Stocking rate animals per hectare	
	Continuous	Rotational	Continuous	Rotational	Continuous	Rotational
1975	9.6	8.5	1.36	1.23	2.00	2.24
1976	9.8	7.9	0.73	1.18	2.42	2.58
1977	8.5	8.7	0.86	1.13	1.74	2.42
1978	8.6	10.3	0.68	0.82	1.71	2.45

ENGLISH

	Conversion ratio lb of forage/lb of beef		Average daily gain lb		Stocking rate animals per acre	
	Continuous	Rotational	Continuous	Rotational	Continuous	Rotational
1975	9.6	8.5	3.0	2.7	0.8	0.9
1976	9.8	7.9	1.6	2.6	0.97	1.0
1977	8.5	8.7	1.9	2.5	0.7	0.97
1978	8.6	10.3	1.5	1.8	0.68	0.98

(Source: P. D. Walton, et al., 1981, J. Range Manage. **34**(1):19–21)

part for the lower animal-weight gains obtained from that treatment. Osuji (1974) estimated that sheep expended 2.27 kJ (0.54 kcal) for each grazing hour for every kilogram of body weight, and values of 1.8 kJ (0.43 kcal)/hr/kg of body weight have also been reported. The extra time spent grazing by the animals on the continuously grazed area represents a significant expenditure of energy.

Changes in the Botanical Composition

The proportions of the plant species that make up a pasture sward are in a constant state of change. The pasture is a dynamic entity which can vary markedly in response to environmental influences, the most important of which is management. A good example of such changes may be found on the University of Alberta Ranch at Kinsella. In 1972, a 100 ha (250 acre) field was sown with a mixture of 1.65 kg/ha (1.5 lb/acre) of alfalfa, 6.7 kg/ha (6 lb/acre) of bromegrass, and 2.75 kg/ha (2.5 lb/acre) of creeping red fescue. The composition by weight was 23% alfalfa, 52% bromegrass, and 25% creeping red fescue, the pasture being uniform over the whole field. Three areas of the field were, respectively, continuously grazed, rotationally grazed, and cut late for hay. The proportion of alfalfa in the continuously grazed area was 12% by 1978. In the same year, there was 42% alfalfa in the rotationally grazed area (see Table 17-9) and 76% alfalfa in the late-cut hay area. In fact, one year's continuous grazing reduced the proportion of alfalfa to about 10%. The reasons for these changes in proportions are evident from data presented in Chapter 8. Under continuous grazing, alfalfa is prefer-

TABLE 17-8
Animal behavior under continuous and rotational grazing, Kinsella 1978

	Mean annual hours per day and standard error	
Grazing system	Grazing	Ruminating
Continuous	10.3 ± 0.70	5.4 ± 0.57
Rotational	7.9 ± 0.61	6.1 ± 0.59
t test significance	1%	n.s.

(Source: P. D. Walton et al., 1981, J. Range Manage. 34(1):19–21)

entially and repeatedly grazed so that carbohydrate root reserves are depleted and the plants eventually die. The rest period provided by rotational grazing permits alfalfa, which replaces carbohydrate reserves slowly, to build up these substances again.

The proportion by weight of creeping red fescue increased under continuous grazing (see Table 17-9). This plant is of low stature, and since much of it is below the bite level of the animal, it is able to regrow quickly and occupy areas from which alfalfa has died out. In contrast, under rotational grazing, the recovery period afforded by the absence of the grazing animals enables the taller species (bromegrass and alfalfa) to regrow and, by competing for light, to suppress the creeping red fescue plants (see Table 17-9). In general, the proportion of smooth bromegrass did not change under either of these two treatments, remaining close to 50%, as at establishment.

Where late haying was practiced, the proportion of bromegrass, by weight, decreased to 8% and that of creeping red fescue to 16%, while the proportion of alfalfa

TABLE 17-9
Percentage mean of species composition by weight under continuous and rotational grazing for an alfalfa–bromegrass–creeping red fescue sward

	Grazing	Alfalfa	Bromegrass	Creeping red fescue
September 1975	Continuous	8.6	41.3	50.1
	Rotational	33.7	27.5	38.8
June 1976	Continuous	10.1	45.6	44.3
	Rotational	26.7	50.9	22.4
June 1977	Continuous	15.5	49.7	34.8
	Rotational	44.4	43.1	12.5
June 1978	Continuous	12.3	54.3	33.4
	Rotational	42.1	45.2	12.7

(Source: P. D. Walton, et al., 1981, J. Range Manage. 34(1):19–21)

increased substantially. The explanation for these changes relates to the reduction in carbohydrate reserves which normally accompanies seed set (see Chapters 7, 8, and 15). For smooth bromegrass, the late-hay cut took place after the seed had formed when the level of carbohydrate reserves was low. Alfalfa did not set seed because of the absence of pollinating insects. Consequently, carbohydrate reserves were high and regrowth was rapid. The alfalfa plants were better able to compete for light, thereby suppressing other species.

The changes in species composition set out in Table 17-9 are reflected in herbage constituents and the digestibility of the herbage (Table 17-10). *In vitro* dry-matter digestibilities decreased successively in June, July, and August for both continuous and rotational grazing. This decrease was more marked for continuous grazing (66.6% to 53.7%) than for rotational (79.7% to 61.3%). Throughout the season, crude protein values were higher for the herbage in the rotationally grazed area than in the continuously grazed area. This difference increased as the grazing season progressed, being 1.7% in June, 5.5% in July, and 8.5% in August.

The levels of calcium, magnesium, and copper were significantly lower in the herbage from the area grazed continuously than in the herbage from the area grazed rotationally. There were no significant differences in the levels of phosphorus found in the two treatments. The higher calcium level found under rotational grazing resulted from a higher proportion of alfalfa found with that type of management. The magnesium level was low for both treatments. Hypomagnesemia has been known to occur at levels below 0.2%. Copper levels of 4 to 5 ppm are considered marginal, so the figures in Table 17-10 are low. Again, higher levels are usually found in legumes than in grasses (Ammerman, 1970, and Sutle, 1976), thus explaining the significantly higher proportion of copper found under rotational grazing.

TABLE 17-10
Seasonal mean values for herbage constituents and digestibility under continuous and rotational grazing, Kinsella, 1978

	Grazing system	
	Continuous	Rotational
Calcium (%)	0.70	1.69[a]
Magnesium (%)	0.07	0.11[a]
Copper (ppm)	4.95	5.98[b]
Phosphorus (%)	0.31	0.29 n.s.
Crude protein (%)	11.40	15.30[c]
In vitro digestibility (%)	58.10	66.40[c]

t test significance level: [a] 0.1%
[b] 5%
[c] 1.0%

(Source: P. D. Walton et al., 1981, J. Range Manage. 34(1):19–21)

Essential Features of Rotational Grazing

What, then, are the main features of rotational grazing? They may be summarized as follows: Rotational grazing gives high forage yields with little wastage from trampling, may reduce parasites for some types of livestock, and permits forage species to be grazed down quickly to the desired height. Because there is a short period of grazing at a high stocking rate, followed by a period of rest, plants are cut or grazed when they have high carbohydrate reserves. Since the grazing period is short, no plant is grazed twice in quick succession. The high stocking rate gives very uniform grazing over the pasture area. Only when stocking rates are high is rotational grazing more productive than continuous grazing.

The disadvantages of rotational grazing are that there is an additional cost for fencing and possibly for water supplies and that it calls for better informed and detailed management. The additional cost of fencing for rotational grazing as compared with the cost for continuous grazing at the University of Alberta Ranch is set out in Table 17-11. With the beef prices prevailing in western Canada at this time, these additional costs would be recovered by the end of the second year. Furthermore, the rotationally grazed pasture, which has more alfalfa, is more productive of green material, which is more digestible, contains more calcium, magnesium, and copper, and gives substantially higher animal-weight gains than a continuously grazed field. In some rotational grazing studies in other parts of North America, gains per animal are reduced, since the cattle are forced to graze down the pasture and eat poor-quality forage. Heavy grazing may be avoided by shortening grazing cycles.

VARIATIONS IN ROTATIONAL GRAZING METHODS

The rotational grazing system described here was selected because of its simplicity. Such systems may be much more complex, as the number of pastures into which the grazed area is divided may be substantially increased. For example, under irrigation, with fertilizers and high production, an electric fence might be used to provide the animals with enough pasture for only one day's grazing; the fencing is then moved every day. Such a program adapts well to areas where sudangrass or sorghum is used for summer pasture. Under these circumstances, it may not be necessary to provide water in the pasture since the animals are given water at milking time. However, moving fences every day makes this system labor intensive.

Another variation involves grazing more than one herd. Dairy cows might be introduced into a pasture for 1 or 2 days, followed by dry cows or young stock to graze off the remaining forage, as used in the Hohenheim system of western Europe. In some cases, selective grazing may be encouraged to give certain production classes of livestock better feed. The "creep grazing" of calves gives them access to high-quality pasture, while excluding the cow with an electric fence or "creep" gate. In parts of Mexico, where irrigation and fertilizers are used together, the number of times the animals move around the rotation may be as high as 25 to 30 times each year.

TABLE 17-11
Additional costs of fencing and water supply for rotational grazing compared to continuous grazing for a 20-hectare field in 1977

	Cost/ha $	Cost/acre $
Materials only		
Fencing	32	12.8
Water supply	35	14.0
	67	26.8
Materials and labor		
Fencing	65	57.2
Water supply	70	61.6
	135	118.8

(Source: P. D. Walton et al., 1981, J. Range Manage. 32(1):19–21)

ZERO GRAZING

Zero grazing (known as soilage, mechanical grazing, or green chopping) may also be regarded as a type of "rotational" grazing. Here the animals are kept in a stockyard; the green material from the pasture is cut and brought in from the field for them daily. In theory, this should lead to a more efficient use of the forage. Since the animal does not expend energy grazing and the pasture is cut uniformly at the optimum time, this method should be more productive. Thus, more animals may be supported per hectare, while the use of the manure produced could be more readily controlled. Where drainage is poor or irrigation is practiced, damage to the soil surface is avoided.

However, in actual fact, such harvesting methods introduce both mechanical and handling losses. Also, the cost of labor (for 7 days per week) and equipment may well mean that this method is uneconomical. Rotational grazing as conducted at Lethbridge, Alberta, is compared with zero grazing in Table 17-12. The points made here are illustrated by these data.

Zero grazing is widely practiced in Europe (Austria, Italy, and Switzerland), as well as in parts of Central and South America (Mexico, Guatemala, Nicaragua, Venezuela, Colombia, Peru, and Uruguay). In these regions, it is usually high-producing milk cows and fattening cattle which make the cutting and carrying of herbage under zero grazing conditions profitable, as such classes of stock are able to consume sufficient herbage for high-level production. In many cases, labor rates are low in the countries where zero grazing is practiced.

TABLE 17-12
Comparison of zero and rotational grazing, Lethbridge, Alberta

	Zero grazing	Rotational grazing	Zero grazing	Rotational grazing
	(kg)		(lb)	
Average animal daily gain	1.1	1.0	2.4	2.2
Gain per hectare	6.83	6.32	6.0	5.6
Production dry matter per hectare (acre)	5,734	6,312	5,103	5,617
Dry matter consumed per kg of animal weight gain	8.1	9.5	8.1	9.5

(Source: Canadian Forage Crop Symposium, 1969)

TABLE 17-13
Total yields (stockpiled plus monthly cutting before stockpiling) of tall fescue for various stockpiling periods, each at four application dates, and rates of 112 kg/ha (100 lb/acre) of nitrogen in Virginia, 1975

Date of application	Stockpiling period			
	June to Dec.	July to Dec.	Aug. to Dec.	Sept. to Dec.
METRIC (kg/ha)				
No nitrogen	2,956	1,996	1,970	1,708
June	4,864	4,728	3,907	3,669
July	4,393	3,905	3,925	3,564
August	3,752	3,846	2,870	3,591
September	3,971	3,820	3,538	3,359
ENGLISH (lb/acre)				
No nitrogen	2,631	1,774	1,753	1,520
June	4,329	4,209	3,477	3,265
July	3,910	3,423	3,493	3,172
August	3,339	3,423	2,554	3,196
September	3,534	3,400	3,149	2,990

(Source: Rayburn et al., 1979, Agron. J. 71)

TABLE 17-14
Crude protein concentration of stockpiled herbage in Kentucky (3-year average)

ACCUMULATION PERIOD	PERCENTAGE OF CRUDE PROTEIN BY COMPONENTS		
	Total herbage	Green herbage	Brown herbage
	Tall fescue		
15 Aug.–1 Oct.	13.6	14.0	6.9
15 Aug.–1 Nov.	11.3	11.6	6.6
15 Aug.–1 Dec.	10.6	11.1	7.7
15 Aug.–8 Feb.	10.1	12.7	8.6
15 Aug.–2 Mar.	10.4	15.6	8.8
	Kentucky bluegrass		
15 Aug.–1 Oct.	17.5	17.9	9.3
15 Aug.–1 Nov.	14.9	15.5	9.8
15 Aug.–1 Dec.	12.6	14.6	9.9
15 Aug.–8 Feb.	11.8	15.2	10.5
15 Aug.–2 Mar.	12.0	17.4	10.5

(Source: Taylor and Templeton, 1978, Agron. J. 68. By permission of The American Society of Agronomy, Inc.)

STOCKPILING

It is possible to accumulate forage in a pasture in order to provide grazing during periods of poor growth. This process is called *stockpiling* and is frequently practiced in the southern United States. Tall fescue, which is widely used in that area, is grazed down in mid-August, fertilized, and allowed to grow until after the first hard frost (Tables 17-13 and 17-14). It is then grazed continuously as long as supplies last. Summer grasses, like bermudagrass, might also be stockpiled for winter use but such material is of decreased quality. The advantage of this system is that there is no danger of overgrazing, since the plant material is already in a dormant state.

CONCLUSION

The first problem faced by the pasture manager is to understand the physiological processes, set out in Chapters 7 and 8, and the factors influencing quality, discussed in Chapter 9. Some of these processes are not completely understood and, even where they are, the ways in which they interact with each other are very complex. Given this knowledge, the pasture manager is then faced with the adaptation of these physiological processes to the environmental conditions in which he or she is farming and the manipulation of these processes to achieve management objectives. Furthermore, pasture management is something of an art, as well as a science. There are those who are quite unacquainted with plant physiology but who, as a result of many years of practical experience, are excellent pasture managers. Those seeking to master pasture management should endeavor to combine the knowledge of science with the wisdom of practical experience.

FURTHER READING

Ammerman, C. B., 1970, Recent development in cobalt and copper in ruminant nutrition, J. Dairy Sci. 53:1097–1107.

Arnold, G. W., 1960, The effect of the quantity and quality of protein available to sheep on their grazing behavior, Aust. J. Agric. Res. 11:1034–1043.

Campbell, J. B., 1961, Continuous versus repeated-seasonal grazing of grass-alfalfa mixtures at Swift Current, Saskatchewan, J. Range Manage. 11:72–77.

Cook, D. A., S. E. Beacom, and W. K. Dawley, 1965, Pasture productivity of two grass–alfalfa mixtures in northwestern Saskatchewan, Can. J. Plant Sci. 45:162–168.

Cooper, M. McG., and D. W. Morris, 1973, Grass Farming, Farming Press Ltd., Ipswich, England, p. 252

Hubbard, W. A., 1961, Rotational grazing studies in western Canada, J. Range Manag. 4:25–29.

Lofgreen, G. P., S. H. Meyer, and J. L. Hull, 1957, Behaviour patterns of sheep and cattle being fed pasture and silage, J. Anim. Sci. 6:773–780.

Osuji, P. O., 1974, The physiology of eating and the energy expenditure of the ruminant at pasture, J. Range Manage. 27:437–443.

Peter, K., 1929, The Hohenheim system, J. Soc. Agron 21:628–633.

Rogler, A. R., 1951, A twenty-five year comparison of continuous and rotational grazing in the Northern Plains, J. Range Manage. 4:35–41.

Sutle, N. F., 1976, The potential toxicity of copper rich animal excreta to sheep, Anim. Production 23:233–241.

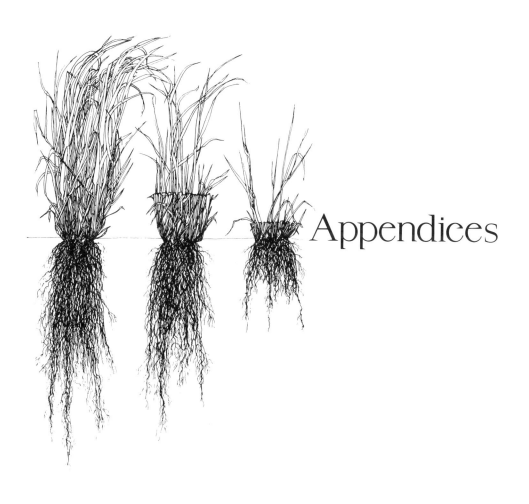

Appendices

Appendix 1   Glossary
Appendix 2   Imperial and metric unit conversion factors

appendix I

GLOSSARY

acuminate: tapering to a point at the end.

acute: sharp-pointed but less so than acuminate.

adaptability: in plants, a modification in the structure or function to fit a changed environment.

aerobic: requiring oxygen to function, as opposed to anaerobic.

agronomy: a science combining crop production and soil management. The word is derived from two Greek words: agros (field) and nomos (to manage).

anaerobic: not requiring oxygen to function.

anther: the pollen-bearing part of the stamen; usually composed of two pollen sacks.

anthesis: stage in floral development when pollen is shed; the period during which the flower is open, and in grasses, the period when the anthers are extended from the glumes.

apical: referring to the apex or main growing tip which is usually the uppermost part.

apical dominance: the inhibition in plants of lateral buds by high levels of auxins produced in the lead shoot or apical meristem.

apomixis: a form of asexual reproduction. Seeds are formed in plants without sexual fertilization.

articulate: provided with joints or made in segments that may be readily separated.

asexual: without sex.

auricle: a clasp-like structure at the base of the leaf blade which aids in support of the leaf to the stem.

awn: the long, extended beak of the lemma or of the outer glume.

axil: the upper angle formed by a leaf or branch with the stem.

axis: the main stem of an inflorescence, especially of a panicle; corresponding to the rachis of a spike.

biennial: of two seasons duration from the time of germination to maturity and death of the plant.

bifid: two-cleft or two-lobed.

bisexual: with flowers containing both stamens and pistils.

blade: the part of the leaf above the sheath.

bloat: in ruminant animals, a condition of excess stomach gas that can result in death, often caused by succulent legumes.

boot stage: the point in cereal plant development when the developing inflorescence is encased in the leaf sheath (boot).

breeder lines: individually selected plants from a polycross, closely checked for uniformity, and bulked to produce the initial seed of a new cultivar.

broadcast: a method of seeding by distributing seed on the soil surface.

bud: the undeveloped state of a stem, branch, or flower.

calyx: the outer or lower circle of bracts of flowers, usually green in color.

cambium: the growing tissue lying between the wood and the bark of a shrub or tree.

carpel: a simple pistil or one section of a compound pistil formed from a modified leaf.

caryopsis: a one-celled fruit with a thin, adherent pericarp or covering.

ciliate: fringed with fine hairs.

companion crop: a crop sown with another; i.e., small grains with which forage crops are sown.

compound leaf: a leaf consisting essentially of a central axis or rachis and several leaflets. It may also possess tendrils and stipules.

convolute: rolled up longitudinally with margins of parts overlapping.

corm: a solid bulb-like enlargement usually at the base of the plant at or below ground level.

corolla: the inner series of floral leaves or petals of a flower, usually white or colored.

cotyledon: a leaf of the embryo within a seed. A storage organ.

coumarin: the bitter flavor substance of sweet clover. Coumarin may occur in lesser amounts in other plants. Coumarin may be converted to toxic substances when spoilage of sweet clover hay or silage occurs.

cover crop: a crop grown to protect the soil from erosion or nutrient leaching.

cross-inoculation groups: symbiotic bacteria are specific for many legumes such as soybeans, but in some cases bacteria will cross-inoculate with several species, e.g., alfalfa and sweet clover.

crown: in plants, the top of a root where buds and new shoots arise.

crown buds: differentiated cells on the top of roots capable of initiating new shoot growth.

culm: the stem or straw of grasses with joints or nodes at intervals.

cultivar (cv.): an inclusive term for lines, varieties, hybrids, or selections of crops. Each cultivar is distinct from other cultivars of the same species.

cytology: the study of individual cells.

decumbent: reclining or lying on the ground, but with the tip ascending.

dehiscence: the opening of a seed pod or anther sack to emit its contents.

determinate: definite cessation or termination of growth of an axis.

diadelphous: stamens in two unequal sets.

dicotyledon: a plant having two cotyledons or seed leaves.

digitate: several members arising from the summit of a support.

dioecious: bearing staminate flowers on one plant and pistillate flowers on another plant as in the date palm.

diploid: having two sets of chromosomes, one from the female and one from the male parent.

direct-seeded forages: seeded without a companion crop.

embryo: the rudimentary plant within the seed.

endosperm: the albumen or starchy part of the seed which serves as food for the young seedling after germination has begun.

epidermis: the outer layer of cells covering the surface of plants.

filament: the stem or pedicel supporting the anther.

flag leaf: the uppermost leaf on a fruiting culm; the leaf immediately below the inflorescence or the seed head.

floret: an individual, small flower such as one of a grass; one of a dense cluster.

fruit: the structure or parts that enclose the seeds.

glabrous: smooth or free from hairs.

glomerate: bunched together in a cluster.

glume: the chaff or bract enclosing the seed of grasses, most commonly referring to one of the two empty bracts of the base of the spikelet.

gynoecium: the pistil or collective carpels of a flower.

haylage: a silage product made from forage with 40% to 60% moisture.

hilum: the scar or mark on the seed where it was attached.

hybrid: a plant or animal which is the result of a cross between two unlike parents.

indehiscent: not dehiscent or not splitting open at maturity.

indeterminate: continuing growth particularly at the apex.

inflorescence: the flowering parts of a plant—a group of flowers on a common axis.

internode: that part of the stem between two nodes.

keel: the two lower united petals of the flower of the pea type; a ridge on the back of the sheath or blade usually along the midrib.

lamina: the blade of a leaf.

lateral: referring to the side.

lax: loose or open, as in a grass panicle.

leaflet: one leaf-like portion of a compound leaf.

lemma: the lower of the two bracts enclosing a grass flower; the flowering glume.

ligule: a collar-like structure projecting upwards, located at the junction of the leaf blade and the leaf sheath.

lodicules: small organs in grasses and cereals located between the ovary and the surrounding glumes. The lodicules swell at the time of anthesis.

monadelphous: stamens united by their filaments into one set.

monocotyledon: a plant having one cotyledon or seed leaf.

monoecious: bearing stamens and pistils on the same plant but in different flowers.

node: the solidified place on the stem which may bear a leaf.

ovate: egg-shaped with the broader end basal.

ovary: that part of the pistil which contains the ovule or ovules.

palea: the chaff or inner bract covering the crease side of the caryopsis of grasses.

panicle: a branching raceme, as in the head of oats.

papilionaceous: pertaining to a particular type of corolla having a standard, wings, and keel, e.g., the flower of a legume.

pedicel: the stalk of a spikelet or of a single flower in a flower cluster.

peduncle: a primary flower stalk supporting an individual flower or flower cluster.

perennial: of three seasons duration or more.

perianth: the calyx and corolla considered together. The term applies most commonly to flowers in which these two parts cannot be readily distinguished.

pericarp: the wall of the ovary. When mature this wall encloses the seed or seeds, as the bran of the wheat or the pod of the pea.

petal: an individual part of the corolla.

petiole: the leaf-stalk or stem by which the leaf is supported.

pistil: the central ovule-bearing organ of a flower, comprising ovary, style, and stigma.

pollen: the grains borne by the anther, containing the male sex cells.

proteins: substances comprised of amino acids and present in all living systems.

pubescent: covered with hairs.

raceme: an elongated indeterminate flower cluster with flowers supported by individual pedicels. The last formed flowers occur at the tip.

racemose: resembling a raceme.

rachilla: the central axis of a spikelet in grasses.

rachis: the central axis of a spike in grasses or the axis of a compound leaf.

radicle: the stem-like shoot of the embryo that grows downward and forms the first roots.

rhizome: an underground stem frequently rooting at the nodes.

scale: a reduced bract or leaf or similar structure usually appressed and dry.

sepal: one separate individual part of the calyx.

sessile: without a stalk as in the spikelets of a spike.

sheath: the lower part of the leaf in grasses enclosing the stem.

silage: feed preserved by the acid-producing action of fermentation.

spike: an unbranched elongated flower cluster with sessile or nearly sessile flowers or spikelets.

spikelet: the unit of inflorescence in grasses consisting of two glumes and one or more florets.

stamen: the male portion of the flower which bears the pollen.

stigma: the sticky or feather-like portion of the pistil to which the pollen grains adhere.

stipule: one of a pair of leaf-like appendages at or near the base of the leaf petiole.

stolon: a basal, horizontal shoot, above the ground, capable of producing roots or new plants at its nodes.

style: the elongated portion of the pistil connecting the ovary and stigma.

tap root: the main root extending vertically downward, other roots being secondary to it without appreciable branching at the crown; a single central root.

tendril: a thread-like leafless shoot occurring on vines or climbing plants and which cling to or coil around objects aiding in the support of the plant.

testa: the outer layer or covering of a seed; developed from the integuments of the ovule, not from the wall of the ovary.

umbel: a determinate, usually convex flower cluster with pedicels arising from a common point.

windrow: a row of cut or uprooted plants raked up or pulled together to dry and to facilitate harvest.

zygomorphic: flowers that can be cut into two equal parts on one plane only.

appendix 2

IMPERIAL AND METRIC UNIT CONVERSION FACTORS

To convert Imperial units (first column) to metric units (third column), multiply by the conversion factor in the second column (e.g., 3 in. × 25 = 75 mm). To convert metric units (third column) to Imperial units (first column), multiply by the conversion factor in the fourth column (e.g., 75 mm × 0.04 = 3 in.).

Imperial unit	Conversion factor to metric unit	Metric unit	Conversion factor to Imperial unit
LINEAR			
inch	25	millimeter (mm)	0.04
foot	30	centimeter (cm)	0.03
yard	0.9	meter (cm)	1.1
mile	1.6	kilometer (km)	0.6
AREA			
square inch	6.5	square centimeter (cm^2)	0.15
square foot	0.09	square meter (cm^2)	11.1
acre	0.40	hectare (ha)	2.5

Imperial unit	Conversion factor to metric unit	Metric unit	Conversion factor to Imperial unit
VOLUME			
cubic inch	16	cubic centimeter (cm³)	0.06
cubic foot	28	cubic decimeter (dm³)	0.04
cubic yard	0.8	cubic meter (m³)	1.25
fluid ounce	28	milliliter (ml)	0.04
pint	0.57	liter (l)	1.75
quart (Imp.)	1.1	liter (l)	0.9
gallon	4.5	liter (l)	0.2
WEIGHT			
ounce	28	gram (g)	0.04
pound	0.45	kilogram (kg)	2.2
short ton (2,000 lb)	0.9	tonne (t)	1.1
PRESSURE			
pounds per square inch	6.9	kilopascal (kPa)	0.15
POWER			
horsepower	746	watt (W)	0.0013
horsepower	0.75	kilowatt (kW)	1.3
TEMPERATURE			
degrees Fahrenheit	(°F − 32) × 0.59	degrees Celsius	(°C − 1.8) + 32
COMMON AGRICULTURAL UNITS			
gallons per acre	11.23	liter per hectare (l/ha)	0.09
quarts per acre	2.8	liter per hectare (l/ha)	0.36
pints per acre	1.4	liter per hectare (l/ha)	0.71
fluid ounces per acre	70	milliliter per hectare (ml/ha)	0.014
tons per acre	2.24	tonnes per hectare (T/ha)	0.45
pounds per acre	1.12	kilograms per hectare (k/ha)	0.89
plants per acre	2.47	plants per hectare (plants/ha)	0.4

(Source: Adapted from various Agriculture Canada publications)

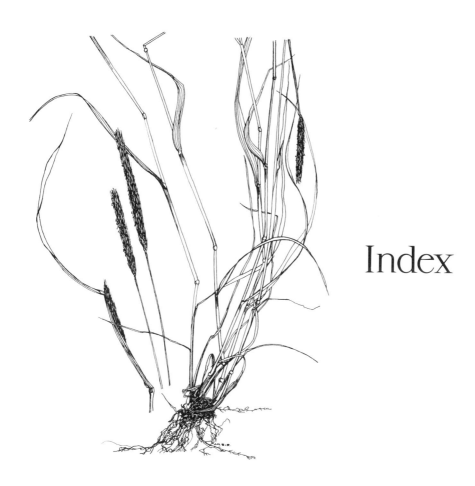

Index

A

Acid-detergent fiber, 251
    determining, 175
Acuminate, definition of, 306
Acute, definition of, 306
Adaptability, definition of, 306
ADF (see Acid-detergent fiber)
Adventitious roots:
    explanation of, 5–6
    illustration of, 3
Aerobic, definition of, 306
Agricultural units, 313
Agronomy, definition of, 306
Alfalfa (*Medicago sativa* L.; *M. falcata* L.; *M. media*), 18, 22, 31, 32, 36, 39, 43, 48, 57, 61, 70, 79–83, 84, 86, 88, 93, 95, 97, 99, 100, 101, 102, 118, 124, 125, 136, 140, 141, 165, 166, 172, 176, 190, 212, 230, 238, 250, 291
    amount 82 days after establishment in winter-formant bahiagrass, 141
    bacterial diseases of, 256
    as bloat causative, 99, 125, 188
    composition and digestibility of fresh and artificially dried, 213
    conditioned and unconditioned stems of, 198, 199
    controlling insect pests of, 254, 255
    cultivars of, 80, 82, 83
    day length in hours and days from initiation of growth to first flower, 265
    decline in carbohydrate reserves associated with repeated defoliation, 163
    digestibility of crude protein in samples dried at different temperatures, 212
    digestion coefficients for herbicide treated and untreated, 251

dry matter yield of newly established mixture of brome and, 19
effect of age on forage quality of, 181, 182
effect of seeding depth on percentage emergence, 131
field cubing of, 210
fungus diseases of, 256, 257
growth rate of flooded and unflooded plants, 83
herbicide for seedling stands of, 251
illustration of, 13, 81
initiating new growth following dormancy, 162–63
moisture requirement and soil tolerance for, 122
nine-year average yield and percentage stand survival following applied potassium, 236
percentage of leaves at various developmental stages for, 180
percentage of nutrient losses when conserved as hay and silage, 228
percentage of sugar content on a fresh-weight basis for, 219
pollination of, 265–67, 269
in preference studies, 278, 279, 280
production and consumption of, 95
proportion of crude fiber when harvested at various stages in growing season, 180
proportion of crude protein when harvested at various stages in growing season, 181
proximate feedstuff analysis of, 174
response to nitrogen applications, 235
Rhizobium inoculation group, 135–36, 237
under rotational grazing, 296–98, 299
saponin content, 187
seed characteristics for, 134, 136
seeding pattern for, 134
seed production for, 262, 274
storage and utilization of available carbohydrates in roots of, 285, 286
virus disease of, 257
windrowing a crop of, 197
Alfalfa aphids, 255
Alfalfa caterpillars (*Colias eurytheme*), 253
Alfalfa weevils (*Hypera postica*), 97, 253, 255
Alkali bee, 269
Alkaloids, 186, 191–92
Alfar (tall wheatgrass cultivar), 61
Allelochemistry, definition of, 185
Alsike clover (*Trifolium hybridum* L.), 36, 61, 86, 88, 125
  as bloat causative, 188
  effect of seeding depth on percentage emergence, 131
  fungus diseases of, 256–57
  illustration of, 13, 87, 88
  moisture requirement and soil tolerance for, 122
  seed characteristics of, 134, 136
  seed production for, 262, 274
  virus disease of, 257
Alta (tall fescue cultivar), 34
Altai wildrye (*Elymus angustus* Trin.), 45, 48
  herbage yield at varying levels of soil salinity, 50
  illustration of, 49
  percentage of emergence from four depths of seeding in greenhouse, 48
  percentage of protein content at four stages of maturity, 51
  seed characteristics for, 135, 137
Amino acids, essential for animal nutrition, 170
Anaerobic, definition of, 306
Anaerobic bacteria, 215, 216
Anid (alfalfa cultivar), 80
Animal behavior, influence of grazing management on, 295–96, 297
Animal manure, total composition of liquid and solid, 239, 240
Animal unit equivalents, 288
Annual forages, 108–20
  carbohydrate reserve levels, 110
  choice between perennial and, 110–11
  crop residues, 114–15
  establishment of summer pastures, 112–13
  establishment of winter pastures, 112
  mixtures, 114, 115
  necessity of a sound grazing system with, 113

reasons for popularity of, 110
uses of, 108–10, 111–13
Annual forages, crop species used as, 109, 115–20
   annual grasses, 109, 117–18
   large grain crops, 109, 116–17
   legumes (*see* Legumes)
   small-grain crops (*see* Small-grain crops)
   succulent fodders (*see* Succulent fodders)
Annual ryegrass (*Lolium multiflorum* Lam.), 43, 45, 70, 262, 291
   illustration of, 46
   seed characteristics for, 135, 137
   wheat yields grown at three densities of, 139, 140
Anthers, 9
   definition of, 306
   illustration of, 10
Anthesis, definition of, 306
Anthranose, 88
Apical, definition of, 307
Apical dome, 4, 5, 6
Apical dominance, definition of, 307
Apomictic, 27, 72
Apomixis, definition of, 307
Armyworms, 252–53
Arrowleaf clover (*Trifolium vesiculosum*), 70, 100–101, 102, 106, 108, 109, 112, 119, 291
   animal weight gains form bermudagrass pastures overseeded with, 112
Articulate, definition of, 307
Asexual, definition of, 307
Aspergillus terreus, 193
Astragalus (genus), 93
Auricle, 4
   definition of, 307
   illustration of, 5, 6
Aurora (alsike clover cultivar), 86
Austrian brome (*see* Smooth bromegrass)
Aveneae (tribe), 148
Avon (orchardgrass cultivar), 41
Awn, definition of, 307
Axil, definition of, 307
Axis, definition of, 307
Azotobacter, 240, 242

B

Bacterial diseases, 256
Bacterial fermentation, process of, 216–17
   butanediol fermentation, 216–17
   mixed acid fermentation, 216, 217
Bahiagrass (*Paspalum notatum* Flügge), 65, 67–69, 72, 74, 101, 141, 291
   average crude protein content and yield harvested at three-month intervals, 67
   average daily gains and beef production from, 74
   dry matter yields for, 286
   illustration of, 68
   number of legumes 82 days after establishment in winter-dormant, 141
   seed characteristics for, 134, 136
Baptisia (genus), 12
Barley, 70, 108, 109, 114, 117, 139
   compared to oats, 117
   dry-matter yield when treated with dry and liquid manure, 239
   proximate feedstuff analysis, 174
   yield of dry matter from mixture grown in same and alternate rows, 115
   yields at different fertility levels grown for forage, 110, 111
Basic slag, 19, 230
Baylor (smooth bromegrass cultivar), 32
Beans, 109, 160
   Rhizobium inoculation group, 135–36
Bentgrasses (genus *Agrostis*), 25, 26
   disease of, 257
Bermudagrass (*Cynodon dactylon* L. Pers.), 65, 66, 67, 69–72, 74, 88, 91, 101, 106, 109, 110, 112, 113, 125, 140, 141, 212, 230, 291
   animal weight gains from overseeded, 112
   average crude protein content and yield harvested at three-month intervals, 67
   controlling winter weeds in, 249
   diseases of, 258, 259
   dry matter yields for, 286
   illustration of, 71

quality characteristics of, 66–67
seed characteristics for, 134, 136
for stockpiling, 302
Biennial, definition of, 307
Bifid, definition of, 307
Big bluegrass (*Poa ampla* Merr.), 27
Big bluestem, 66
Birdsfoot trefoil (*Lotus corniculatus*), 83–84, 95, 99, 165, 186
herbicide for seedling stands of, 251
illustration of, 13, 84, 85
moisture requirement and soil tolerance for, 122
no cases of bloat, 84, 188
in preference studies, 278, 279
seed characteristics for, 134, 136
Bisexual, definition of, 307
Blade, definition of, 307
Bleeding disease of cattle, 99, 189
Blister beetles (*Epicauta murina*), 253, 254
Bloat, 187–89
causes of, 187–88
definition of, 307
description of, 187
prevention of, 189
Bluebunch (*see* Idaho fescue)
Bluegrasses (genus *Poa*), 25, 27–29
Blue panicgrass, 66
Bluestems, 72, 263
diseases of, 259
Boot stage, definition of, 307
Boreal (creeping red fescue cultivar), 36
Bottom grass (*see* Creeping red fescue)
Bound water, definition of, 158
Breeder lines, definition of, 307
Broadcast seeding, 132, 133, 307
Bromegrass, 22, 25, 36, 43, 150, 176, 291
diseases of, 259, 260
dry matter yields of newly established mixture of alfalfa and, 119
effect of ammonium nitrate fertilizer on first-cut yields of hay, 231
effect of seeding depth on percentage emergence, 131
moisture requirement and soil tolerance for, 122

percentage of emergence from four depths of seeding in greenhouse, 48
in preference studies, 278, 279
proximate feedstuff analysis, 174
response to nitrogen applications, 232
sod-bound condition, 232
use of TAC, 147
(*see also* Smooth bromegrass)
Bromus (genus), 149
Broomsedge, 72
Buckwheat, 120
Bud, definition of, 307
Buffalograss, 77, 128, 263 (*see also* Saint Augustinegrass)
Buffelgrass, 66, 263
Bunchgrass, 34

C

Cabbage, 120
Callil (bermudagrass cultivar), 72
Calyx, definition of, 307
Cambium, definition of, 307
Canada bluegrass (*Poa compressa* L.), 27
seed characteristics of, 134, 136
Canada Seeds Act, 273
Canada wildrye (*Elymus canadensis* L.), 45
Canadian Seed Growers Association, 121
Carbohydrate substances:
nonstructural, 170 (*see also* Forage crop growth, grasses, carbohydrate substances used in)
structural, 170–71, 181
Carotenoids, 172
Carpel, definition of, 307
Carpetgrass (*Axonopus affinis* Poir), 65, 66, 72, 91
diseases of, 258
Caryopsis, definition of, 11, 307
Castor (reed canarygrass cultivar), 43
Caterpillars, 252, 253, 254–55
Cattail, uses of, 108, 109
Champ (timothy cultivar), 52
Chewing's fescue (*Festuca rubra* sub. sp. *commutata*), 36

Chief (intermediate wheatgrass cultivar), 57
Chinch bugs, 253
Chinook (orchardgrass cultivar), 41
Chloridoideae, light sensitivity of, 127
Cicer milkvetch (*Astragalus cicer* L.), 93, 95
    illustration of, 92
    no cases of bloat, 95
    production and consumption of, 95
    seed characteristics of, 134, 136
Ciliate, definition of, 307
Climax (timothy cultivar), 52
Clostridium, 242
Clover, 12, 18, 20
    illustration of, 13
Clover root curculio (*Sitona hispidula*), 254
Coastal (bermudagrass cultivar), 70, 72, 100, 101
    pH value tolerance, 238
    production rate and protein concentration in, 142
    response to nitrogen applications, 231, 234
    response to sulfur applications, 234
Coastcross-1 (bermudagrass cultivar), 70, 72
Cocksfoot (*see* Orchardgrass)
Colonial bent (*Agrostis tenuis*), 25
    seed characteristics for, 134, 136
Commercial seed production practices, 274–75
Common vetch (*Vicia sativa* L.), 263
    illustration of, 98
Companion crops, 139–40
    definition of, 307
Compound leaf, definition of, 308
Conditioning herbage, process of, 198
Convolute, definition of, 308
Cool-season perennial grasses, 25–65, 140
    compared to warm-season perennial grasses, 65–66
    floral induction requirements of, 25
    herbage yield of six grasses at varying levels of soil salinity, 50
    main carboxylating enzyme for, 25
    nitrogen fertilization of, 231, 232
    percentage of emergence of four grasses from four depths of seeding in greenhouse, 48
    percentage of protein content for four grasses at four stages of maturity, 51
    quality characteristics of, 66
    response to potassium fertilization, 233
    summer dormancy, 158
    tolerance of soil acidity, 238
    winter formancy in, 157
Cool-season perennial grasses, species of:
    bentgrasses, 25, 26, 257
    bluegrasses, 25, 27–29
    fescue, 32–36, 37, 70, 80, 101, 106
    meadow foxtail, 36, 38–39, 66, 194
    orchardgrass (*see* Orchardgrass)
    reed canarygrass (*see* Reed canarygrass)
    ryegrasses (*see* Ryegrasses)
    smooth bromegrass (*see* Smooth bromegrass)
    timothy (*see* Timothy)
    wheatgrasses, 45, 52–53, 55–65, 259
    wildrye, 45, 47–50, 259
Corm, definition of, 308
Corn (*Zea mays* L.), 82, 108, 109, 114, 116, 119, 148, 160
    disease of, 258–59
    dry matter yield of, 119
    percentage of sugar content on a fresh-weight basis, 219
    yield and digestibility values from, 116
Corolla, definition of, 308
Cotyledon, definition of, 308
Couch (*see* Bermudagrass)
Coumarin, 186
    conversion to dicamarol, 99, 188, 189
    definition of, 308
Coumestan compounds, 106
Coumestrol, 190
Cover crop, definition of, 308
Cowgrass (*see* Red clover)
Cowpeas, 100, 109, 116, 117, 119, 291
    disease of, 258
    Rhizobium inoculation group, 135–36
Crabgrass (*see* Saint Augustinegrass)
Creeping bent (*Agrostis palustris*), 25
    seed characteristics of, 134, 136

322 / Index

Creeping foxtail, percentage of nitrogen recovered as influenced by rate and frequency of application, 233
Creeping red fescue (*Festuca rubra* sub. sp. *rubra*), 36, 124, 262
    dry matter production from, 57
    germination of, 127
    illustration of, 37
    moisture requirement and soil tolerance for, 122
    physiological stages of, 264
    in preference studies, 278, 279
    under rotational grazing, 296–98
    seed characteristics for, 134, 136
Crested wheatgrass (*Agropyron cristatum* L.; *Agropyron desertorum*), 22, 36, 45, 52–53, 55–57, 61, 97, 194, 290
    effect of seeding depth on percentage emergence, 131
    for heavy grazing, 289
    illustration of, 55, 56
    moisture requirement and soil tolerance for, 122
    in preference studies, 278, 279
    seed characteristics for, 135, 137
    seed production for, 274
Crimson clover (*Trifolium nicaenatum*), 70, 100, 101, 106, 108, 109, 112, 119, 166, 238, 291
    animal weight gains from bermudagrass pastures overseeded with, 112
    fungus disease of, 257
    stem and crown rots, 258
    virus disease of, 257
Critical leaf area, definition of, 150
Crop residues, processing with sodium hydroxide, 115
Cross-inoculation groups, definition of, 308
Crown, definition of, 162, 308
Crown buds, 197
    definition of, 308
Crown vetch, seed characteristics for, 135, 137
Crude fiber, 179, 180, 181, 228, 280
    determining, 174
Crude protein, 179, 181, 182, 199, 212, 228, 251, 280, 302
    determining, 173–74

Cryobiology, definition of, 158
Culm, 6
    definition of, 5, 308
Cultipacker, definition of, 132
Cultivar, definition of, 308
Cutworms, 252
Cyanogenetic glucosides, 186
Cytology, definition of, 308

D

Dactylis (genus), 149
Dalapon, 141
Dallisgrass (*Paspalum dilatatum* Poir), 65, 66, 69, 72, 91, 141, 159, 263, 291
    diseases of, 258
    dry matter yields for, 286
    illustration of, 73
    response to nitrogen applications, 231
    seed characteristics for, 134, 136
Dawson (red fescue cultivar), 36
Decumbent, definition of, 308
Defoliation, 154–57, 289
Dehiscence, definition of, 308
Dehydration, 183, 196, 226
    process of, 212–13
Delta (Kentucky bluegrass cultivar), 27
DES (*see* Diethylstilbestrol)
Determinate, definition of, 308
Diadelphous, definition of, 308
Dicotyledon, definition of, 308
Dicoumarol, 99, 188, 189
Diethylstilbestrol, 190
Digestible energy, definition of, 176
Digitate, definition of, 308
Digitgrass (*Digitaria decumbens* Stent.), 74
    average crude protein content and yield harvested at three-month intervals, 67
    average daily gains and beef production from, 74
Dioecious, definition of, 308
Diploid, definition of, 308
Direct-seeded forages, definition of, 308
Dixie (crimson clover cultivar), 101

Druchamp (wheat cultivar), grain yields grown at three ryegrass densities, 140
Dry-matter loss, 226, 227, 228

E

Early bluegrass (*Poa cusickii* Vasey), 27
18S protein, 188
Embryo, definition of, 308
Empire (birdsfoot trefoil cultivar), 84
Endosperm, 12
    definition of, 11, 308
    importance in root growth, 130
English wild white clover, 91
Ensilage (*see* Silage-making)
Epidermis, definition of, 309
Ergot fungus, 72, 248
Eski (sainfoin cultivar), 95, 97
Establishment, 121–43
    choice of crops, 123
    companion crops, 139–40, 307
    grassland renovation, 140–43
    importance of germination, 126–28
    importance of seed quality for, 125–26
    making a forage mixture, 125
    seedbed and seeding depth, 130–32
    seeding equipment, 132
    seeding pattern, 133–34
    seedling vigor (*see* Seedling vigor)
    seed rates, 133, 134, 135, 136, 137
    seed treatment, 134–39
    selection of mixtures, 123–25
    time of planting, 132–33
Ether extractives, 175, 179, 228

F

Fairway (crested wheatgrass cultivar), 57
Fawn (tall fescue cultivar), 34
Federal Seed Act, 273
Fertilizers, 229–43
    on grasses, 230–34
    on grass-legume mixtures, 237–38
    importance of, 229–30
    introduction of, 19, 230
    on legumes, 235–36
    nitrogen balance in pastures, 240–43
Fertilizers, plant and animal waste, 238–40
    total composition of liquid and solid animal manure, 239, 240
Fescue foot:
    causes of, 192–93
    explanation of, 192
Fescues (genus *Festuca*), 32–36, 37, 70, 88, 101, 106
Festuceae (tribe), 29, 43, 52, 149
Festucoideae, light sensitivity of, 127
Filament, definition of, 309
Flag leaf, definition of, 309
Flavonoids, 106, 186, 190
Flax, 139
Floret, 9
    definition of, 309
    illustration of, 10
Forage crop diseases, 256–61
    control methods, 260–61
    of grasses, 258–60
    of legumes, 256–58
Forage crop growth, grasses, 144–60, 289
    carbohydrate reserves management, 149–50
    carbohydrate reserves necessary for, 146–48, 158
    carbohydrate substances used in, 145, 146
    defoliation of a tiller, 154–57
    growth and regrowth, 150–53
    mineral nutrients required for, 144, 145
    photosynthesis for, 145, 146–47, 149, 150, 159
    plant stress, 158–60
    relationships between storage substances and species, 148–49
    respiration for, 145, 147, 148, 149, 150, 155, 159
    seasonal changes in growth rates, 155–57
    storage regions, 148
    summer dormancy, 158
    winter dormancy, 157
Forage crop growth, legumes, 161–66, 289
    canopy structure, 165
    growth responses, 164–66
    influence of light, 164–65

leafage, 165–66
respiration, 166
root system, 164
shoot growth, 162–63
soil moisture, 166
Forage crop pests, 244–55
Forage looper (*Caenurgina erechtea*), 253
Forage plant breeding practices, 270–73
Forage production, economic efficiency of, 284
Forage quality, 22, 168–83
Forage quality, causes for variation in, 178–82
    climate, 182
    fertilization, 182
    management of yield and quality, 183, 199
    relation between leaf-to-stem ratio and plant age, 179–82
    time of harvesting, 179
Forage quality, chemical analysis, 168, 169–73
    carbohydrates including pectin, 170–72
    minerals, 172–73
    nine principal groups of chemical constituents in forage, 169
    nitrogenous compounds, 169–70
    nonprotein nitrogenous compounds, 170
Forage quality, effect of chopping, grinding, and pelleting on feed intake and liveweight gains, 177, 178
Forage quality, measurement factors, 168–78
    animal intake, 176–78
    chemical analysis (*see* Forage quality, chemical analysis)
    digestibility, 168, 176, 177, 181
    energy intake, 168, 175–76
    proximate feedstuff analysis, 168, 173–75
Forages:
    amount of seed needed to establish new stand, 133, 134, 135, 136, 137
    annual (*see* Annual forages)
    companion crops, 139–40, 307
    critical leaf area index for, 150, 151
    definition of, 2
    effect of seeding depth on percentage emergence, 130, 131
    importance of germination, 126–28
    importance of seed quality, 125–26
    nine principal groups of chemical constituents in, 169

    objectives for making a mixture, 125
    peak production at different times in growing season, 123
    planting time for, 132–33
    reasons for growing in mixtures, 123–25
    relative moisture requirement and soil tolerance for, 122
    seedbed, 130
    seeding equipment, 132
    seeding pattern for, 133–34
    seedling vigor (*see* Seedling vigor)
    seed treatment to aid establishment and improve yield, 134–39
    sowing seeds, 132
Forages, antiquality factors, 185–94, 247–48
    alkaloids, 186, 191–93
    coumarin, 99, 188, 189
    cyanogenetic glucosides, 186
    18S protein, 188
    explanation of, 185
    flavonoids, 186, 190
    low levels of magnesium, 194
    nitrate poisoning, 193
    saponins, 186, 187
    tannins, 186, 188–89, 190
Forages, families of:
    grasses (*see* Grasses)
    legumes (*see* Legumes)
Forage seed production, 262–75
Forage seed production, pollination mechanisms:
    in grasses, 263–64
    leaf-cutter bees, 265, 266, 267, 268, 269
    in legumes, 265–68
Forage seed trade organizations, 270–75
    commercial seed production practices, 274–75
    harvesting forage seed, 275
    plant-breeding practices, 270–73
    post harvest residue management, 274–75
    seed multiplication, 273–74
    weed control, 274
Forage storage, 195–96
    dry matter and nutrient loss in, 226–28
    dry systems (*see* Haymaking)
    field and storage losses, 226, 227
    silage (*see* Silage-making)

Forage yield variation, 284–87
    solutions to, 288
Formaldehyde, 223
Formic acid, 222, 223
Foxtail millet, 113
    diseases of, 258–59
Frontier (reed canarygrass cultivar), 43
Fructosans, 145, 148, 149, 169
Fructose, 145, 149
Fruit, definition of, 309
Fungus diseases, 256–57

G

Gama dulce (*see* Bahiagrass)
Gangrenous ergotism, 192, 193
Genetic homeostasis, 272–73
Genetic purity, 121, 125–26
Gengibrillo (*see* Bahiagrass)
Germination, 121, 126–28
    definition of, 126
    procedures to enhance process for forage seeds, 127–28
Gibberellic acid, 127
Glabrous, definition of, 309
Glomerate, definition of, 309
Glucose polymer:
    amylopectin, 148
    amylose, 145, 148
Glume, 9
    definition of, 309
    illustration of, 10
Glyphosate, 141, 142
Goar (tall fescue cultivar), 34
Grain sorghums:
    grazing, 117
    grinding, 117
    as silage, 116–17
Gramma de agua (*see* Dallisgrass)
Grasses, 3–11
    annuals as forage, 117–18
    cool-season perennial (*see* Cool-season perennial grasses)
    definition of, 3
    delineation of growing areas in the U.S., 24
    diseases of, 258–60
    germination in, 126, 127
    as gross nitrogen feeders, 230
    herbage yield of six grasses at varying levels of soil salinity, 50
    illustration of elongated grass plant, 6
    illustration of nonelongated grass plant, 3
    illustration of types of auricles, 5
    illustration of types of ligules, 4
    inflorescence of (*see* Grass inflorescence)
    mixing legumes with, 19
    percentage of emergence of four grasses from four depths of seeding in greenhouse, 48
    percentage of protein content for four grasses at four stages of maturity, 51
    pollination mechanisms, 263–64
    principal mineral constituents of, 173
    relative moisture requirement and soil tolerance for, 122
    seed of, 11
    seed characteristics for legumes and, 134
    shoot system of, 4–5, 6, 7, 8
    stem types, 5, 7, 8
    vegetative stage of, 3, 4
Grasses, nutrient requirements of, 230–34
    mixed with legumes, 237–38
    nitrogen, 230–33
    phosphorus, 233, 234
    potassium, 233–34
    sulfur, 234
Grasses, root system of, 5–6
    adventitious roots, 3, 5–6
    seminal roots, 3, 5, 11
Grasshoppers, 124–25
Grass inflorescence, 7, 9–10, 263–64
    components of, 9–10
    illustration of, 9, 10
    reasons for different appearances of, 9
    types of, 7
Grassland renovation, 140–43
    summary of animal preference on improved and unimproved pasture, 142
Grasslands:
    adaptations for survival under grazing, 17–18
    extent over earth's surface, 17

326 / Index

fulfilling man's need in two ways, 18
productivity of, 1–2
supporting livestock industry, 2
Grasslands, historic perspective on, 16–22
    beginning of farming, 18–19
    farming in Europe, 19
    farming in North America, 20–22
    geology of, 16–18
Grass tetany, 194
Grazing behavior, 280–81
Grazing systems, kinds of:
    alternate, 290–91
    complementary, 290, 292
    deferred, 290, 292
    rotational (*see* Rotational grazing)
    zero, 300–301
Greenleaf (pubescent wheatgrass cultivar), 60
Green stipa hay, 177
Gros chiendent (*see* Bermudagrass)
Growth-regulator weed killers, 251
Guineagrass, pH value tolerance, 238
Gynoecium, definition of, 309

H

Hairy vetch (*Vicia villosa*), 99, 263
    seed characteristics for, 135, 137
Harden, definition of, 133
Haylage, definition of, 309
Haymaking, 196–213
    dehydration process, 212–13
    losses under fair conditions, 211
    percentage of digestible protein lost due to rain while, 211
Haymaking, methods of:
    for baled hay, 199, 203–8
    barn dying, 208–9
    for chopped hay, 199, 202–3
    for long hay, 199–202
    loose hay systems, 209, 210
    for shredded hay, 199, 203
    use of hay additives in, 210–11
    wafered and pelleted, 199, 209–10
Haymaking, principles of, 196–99
    avoidance of fermentation, 198–99
    conditioning, 198
    timeliness of harvesting, 198, 199, 200
    windrowing, 196–97, 200, 201, 203, 223, 311
Hay stage, definition of, 125
Herbicides, use of, 131–32, 141, 142, 250–52
    nonselective, 250
    selective, 250–51
Herd behavior, 281–82
Highlight (red fescue cultivar), 36
Hilum, definition of, 309
Hordeae (tribe), 52, 148
Hormone weed killers, 251
Hubam (white sweetclover cultivar), 99
Hungarian brome (*see* Smooth bromegrass)
Hybrid, definition of, 309
Hydrocyanic acid, 186
Hypomagnesemia, 194
Hyslop (wheat cultivar), grain yields grown at three ryegrass densities, 140

I

Idaho fescue (*Festuca idahoensis* Elmer), 32
Imperial units, conversion to metric, 312–13
Inbreeding depression, definition of, 15
Indehiscent, definition of, 309
Indeterminate, definition of, 309
Indiangrass, 66
Inflorescence, definition of, 309. (*See also* Grass inflorescence.)
Insect pests, 252–55
    classification of, 252–54
    control methods, 254–55
Intermediate wheatgrass (*Agropyron intermedium*), 52, 57, 60
    dry matter production from, 57
    fall utilization for animals given free choice, 60
    illustration of, 9, 58
    moisture requirement and soil tolerance for, 122
    in preference studies, 278, 279
    seed characteristics for, 135, 137

Internodes
    definition of, 309
    in grasses, 8, 45
    in legumes, 12
Italian ryegrass (*Lolium multiflorum* Lam.), 43, 45, 70, 88, 109
    germination of, 127
    illustration of, 46
    seed characteristics for, 135, 137

J

Johnsongrass, 65, 69, 74, 118, 142, 186, 248, 291
    diseases of, 258, 259
    illustration of, 75
    seed characteristics for, 134, 136
June bug, 254
Junegrass (*see* Kentucky bluegrass)

K

Kale, 109, 120
    dry matter yield of, 119
Kay (orchardgrass cultivar), 41
Keel, definition of, 309
Kenhy (tall fescue cultivar), 34
Kenon (Kentucky bluegrass cultivar), 27
Kenstar (red clover cultivar), 88, 91
Kentucky (crimson clover cultivar), 101
Kentucky bluegrass (*Poa pratensis* L.), 27, 36, 65, 88, 91, 140, 149, 194, 262, 263, 291
    diseases of, 259, 260
    effect of seeding depth on percentage emergence, 131
    for heavy grazing, 289
    illustration of, 28
    number of shoots and rhizomes per sq. meter, 29
    pH value tolerance, 238
    potassium requirement of, 234
    seed characteristics of, 134, 136
Kentucky 31 (tall fescue cultivar), 34
Kenwell (tall fescue cultivar), 34

Kobe lespedeza (*Lespedeza striata*), 102
Korean lespedeza (*Lespedeza stipulacea*), 102
Kudzu (*Pueraria lobata* Willd.), 100, 101–2, 238, 291
    illustration of, 103
    disease of, 258

L

Ladino clover, 39, 84, 91, 93, 100, 159, 166, 212, 230
    amount 82 days after establishment in winter-dormant bahiagrass, 141
    effect of applied nitrogen on nitrogen fixation in, 235
    effect of seeding depth on percentage emergence, 131
Lady beetles (*Hippodamia convergens*), 255
Lahoma (sudangrass cultivar), 118
LAI (*see* Leaf area index)
Lamina, definition of, 309
Large-grain crops, 109, 116–17
Lateral, definition of, 309
Lax, definition of, 309
Leaf and stem crops, 119, 120
Leaf area density, definition of, 165
Leaf area index, 151, 165
    definition of, 150, 165
Leaf-cutter bees (*Megachile rotundata* Fabricus), 265, 266, 267, 268, 269
Leaflet, definition of, 309
Leaf sheaths, 6
    definition of, 4
    illustration of, 3
Leafspot fungi, 190
Leaf-to-stem ratio, 179, 181
Legumes, 11, 12–15, 79–106, 109, 119
    crude protein content of, 12
    developing root nodules with Rhizobium, 12
    diseases of, 256–58
    inflorescence, 12, 15
    mixing with grasses, 19
    parts of flower, 12, 14
    pollination mechanisms, 265–68

principal mineral constituents of, 173
roots of, 12
seed characteristics for grasses and, 134
stems and leaves of, 12, 13
warm-climate, 99–106, 109, 140
Legumes, nutrient requirements of, 235–36
mixed with grass, 237–38
Legumes, in temperate areas, 79–99, 109
relative moisture requirement and soil tolerance for, 122
Lemma, 9, 11
definition of, 309
illustration of, 10
Leo (Birdsfoot trefoil cultivar), 84
Lespedeza, 70, 72, 100, 104, 108, 109, 113, 119, 125, 140, 262, 263
diseases of, 258
seed characteristics for, 134, 136
Lignin, 171, 174, 175, 181
Ligules, 4
definition of, 309
illustration of, 4, 6
Lime application, 237–38
Lipids, 171
Little bluestem, 66
Lodicules, 11
definition of, 9, 309
illustration of, 10
Louisiana S-1 (white clover cultivar), 93
Louisiana white clover, 91
Lucerne (*see* Alfalfa)
Lupines, *Rhizobium* inoculation group, 135–36
Lutana (cicer milkvetch cultivar), 95

M

Madrid (yellow sweetclover cultivar), 99
Maitland (birdsfoot trefoil cultivar), 84
Mandan (Canada wildrye cultivar), 45
Mandan 759 (pubescent wheatgrass cultivar), 60
Mangoes, 119
Mass selection breeding technique, 270–72
May beetle, 254

Meadow clover (*see* Red clover)
Meadow fescue (*Festuca elatior* L.), 32, 34–36
germination of, 127
illustration of, 35
in preference studies, 280
seed characteristics for, 134, 136
Meadow foxtail (*Alopecurus pratensis* L.), 36, 39, 66, 194
illustration of, 38
Meadow grass (*Poa trivialis* L.), 27
Meadow spittlebugs (*Philaenus supmarius*), 253, 255
Mechanical cultivation, 130, 131
Medicago (genus), 188–89
Medics (*see* Alfalfa)
Melrose (sainfoin cultivar), 97
Merion (Kentucky bluegrass cultivar), 27
Metabolizable energy, definition of, 176
Metric units, conversion to imperial, 312–13
Midland (bermudagrass cultivar), 70, 72
Migratory grasshopper (Melanoplus sanguinipes), 252, 254
Millet, 117, 125
Missouri (tall fescue cultivar), 34, 69
Mites, 253, 254
Molasses, 219, 220, 222
Monadelphous, definition of, 309
Monocotyledon, definition of, 309
Monoecious, definition of, 309
Mt. Barker (subterranean clover cultivar), 106
Mulches, use of, 132
Mungbeans, 119
Mustard, 120
Mutton bluegrass (*Poa fendleriana* Vasey), 27
Mycoplasma diseases, 257

N

Napiergrass, 65, 67
pH value tolerance, 238
New York wild white clover, 91
NFE (*see* Nitrogen-free extracts)
Nitrate poisoning, 193, 233
Nitrogen deficiencies, 230

Nitrogen fertilizers:
    applied to grass-legume mixture, 237
    applied for grass seed production, 274
Nitrogen fertilizers, applied to grasses:
    danger of nitrate poisoning, 193, 233
    effect of ammonium nitrate on first-cut yields of bromegrass hay, 231
    percentage recovered from three grasses influenced by rate and frequency of, 233
    protein content of fertilized and unfertilized bromegrass, 232
Nitrogen fertilizers, applied to legumes, 235
    effect of applied nitrogen on nitrogen fixation in established ladino clover, 235
    response of alfalfa to, 235
Nitrogen fixation:
    amount of, 240–42
    biochemistry of, 138
    factors influencing, 139, 235, 237
Nitrogen-free extracts, 174, 175, 179, 228
Nitrogen, transformations in grassland ecosystem, 241
Node, definition of, 309
Nolin (white clover cultivar), 93
Nordan (crested wheatgrass cultivar), 57
Norlea (ryegrass cultivar), 45
Nugaines (wheat cultivar), grain yields grown at three ryegrass densities, 140
Nurse crops (*see* Companion crops)
Nutrient loss, 226–28

O

Oahe (intermediate wheatgrass cultivar), 57
Oats, 51, 70, 108, 112, 114, 117, 139
    correct height for grazing, 111
    mean yields of dry matter and crude protein for, 114
    percentage of sugar content on a fresh-weight basis for, 219
    proximate feedstuff analysis, 174
    seed characteristics for, 135, 137
    yield of dry matter from oats and peas grown in same and alternate rows, 115
    yields at different fertility levels grown for forage, 110, 111
Orbit (tall wheatgrass cultivar), 61
Orchardgrass (*Dactylis glomerata* L.), 27, 32, 39–41, 65, 66, 83, 88, 100, 125, 148, 149, 150, 153, 166, 212, 230, 264, 291
    diseases of, 259, 260
    germination of, 127
    illustration of, 9, 40
    moisture requirement and soil tolerance for, 122
    pH value tolerance, 238
    potassium requirement of, 234
    in preference studies, 280
    protein production per unit acre of land, 52
    proximate feeding stuff analysis results for, 41
    seed characteristics for, 135, 137
    weedy species associated with, 246
    yield in monoculture and in mixture with reed canarygrass, 123
Organic acids, 171
Ovary, definition of, 310
Ovate, definition of, 310
Oxley (cicer milkvetch cultivar), 95

P

Pacific (ryegrass cultivar), 45
Palatability, 176, 276–80
    determination of preference, 277–80
    explanation of, 276–77
Palea, 9, 11
    definition of, 310
    illustration of, 10
Pangolagrass (*Digitaria decumbens* Stent.), 65, 67, 74
    pH value tolerance, 238
Panicle, definition of, 310
Panicoideae, light sensitivity of, 127
Panmixis plots, 272

Papilionaceous, definition of, 310
Paraquat, 141
Parkway (crested wheatgrass cultivar), 57
Paspalum (genus), 67
    germination of, 127
Paspalum dilatatum, 66
Pasto bermuda (see Bermudagrass)
Pasto horqueta (see Bahiagrass)
Pasto miel (see Dallisgrass)
Pasture management, 283–303
    determination of stocking rates, 288
    economic importance of, 283, 284
    forage yield variation, 284–87, 288
    goals of pasture manager, 284, 290, 302–3
    plant characters capable of withstanding frequent and close grazing, 289
    undergrazing and overgrazing, 289–90
Pastures, types of, 290
Pea aphids (*Acyrthosiphon pisum* Haris), 187, 253, 254, 255
Peanuts, 108, 119
Pearl millet (*Pennisetum glaucum*), 108, 109, 112–13, 117
    diseases of, 258–59
Peas, 109, 119, 139
    mean yields of dry matter and crude protein for oats and, 114
    Rhizobium inoculation group, 135–36, 138
    yield of dry matter from peas and oats grown in same or alternate rows, 115
Pedicel, definition of, 310
Peduncle, definition of, 310
Pelleting, definition of, 139
Penncross (bentgrass cultivar), 25
Pensacola (bahiagrass cultivar), 69
    response to nitrogen applications, 231
Perennial, definition of, 310
Perennial ryegrass (*Lolium perenne* L.), 19, 34, 43–45, 66, 86, 212, 262, 264
    illustration of, 44
    pH value tolerance, 238
    seed characteristics for, 135, 137
    weedy species associated with, 246
Perianth, definition of, 310

Pericarp, 11
    definition of, 11, 310
Petal, definition of, 310
Petiole, definition of, 310
Phalaris, average annual performance of steers on, 34
Phalaris staggers, 192
Phaseolus, 238
Phleum bertolonii, 50
Phleum commutatum, 50
Phleum nodosum, 50
Phosphorus fertilizers, applied to grasses, 19, 230, 233
    effect of annual applications of phosphate on old brome grass stand, 234
Phosphorus fertilizers, applied to legumes, 235
    effect of alfalfa on available phosphorus at varying soil depths, 236
Photosynthesis:
    in grass-legume mixtures, 237
    in grasses, 145, 146–47, 149, 150, 159
    in legumes, 164–66, 236
Picloram, 141, 142
Pistil, definition of, 310
Plant Variety Protection Act, 274
Plowing, importance of, 130
Poa (genus), 158
Poa scabrella, 158
Poisonous weeds, 247–48
Polara (white sweetclover cultivar), 99
Pollen, definition of, 310
Polycross plots, 272
Potassium carbonate, 211
Potassium fertilizers:
    applied to grasses, 233–34
    for grass-legume mixture, 237
Potassium fertilizers, applied to legumes, 235–36
    nine-year average alfalfa yield and percentage stand survival using, 236
Potassium nitrate, 127
Potato leafhoppers (*Empoasca fabae*), 253, 254, 255
Prairieland (altai wildrye variety), 48
Prickly pear (*Opuntia* spp.), 249–50
Primar (slender wheatgrass cultivar), 61

Propionic acid, 210
Prostrate (Dallisgrass cultivar), 72
Proteins, definition of, 310
Proximate feedstuff analysis, 173–75
Prussic acid, 186
Pubescent, definition of, 310
Pubescent wheatgrass (*Agropyron trichophorum*), 52, 57, 60
    illustration of, 59
    moisture requirement and soil tolerance for, 122
    seed characteristics for, 135, 137
    spike of, 60
Purple clover (*see* Red clover)
Pyruvic acid, 216, 217

Q

Quackgrass, 61

R

Raceme, explanation of, 7, 310
Racemose, definition of, 310
Rachilla, definition of, 310
Rachis, 7, 9
    definition of, 310
Radicle, definition of, 310
Rapeseed, 109, 120
    yield of dry matter from rapeseed and oats grown in same and alternate rows, 115
Red clover (*Trifolium pratense* L.), 19, 70, 86, 88–89, 91, 101, 112, 125, 165, 166, 238
    amount 82 days after establishment in winter-dormant bahiagrass, 141
    as bloat causative, 188
    effect of seeding depth on percentage emergence, 131
    fungus diseases of, 256–57
    illustration of, 13, 89
    moisture requirement and soil tolerance for, 122
    proximate feedstuff analysis of, 174
    seed characteristics of, 134, 136
    seed production for, 262, 274
    types of, 86
    virus disease of, 257
    weedy species associated with, 246
Red fescue (*Festuca rubra* L.), 5, 32, 36, 264
    disease of, 257
Redland (red clover cultivar), 91
Redtop (*Agrostis alba*), 25
    dry matter production from, 57
    effect of seeding depth on percentage emergence, 131
    fall utilization for animals given free choice, 60
    germination of, 127
    illustration of, 26
    in preference studies, 278
    seed characteristics of, 135, 137
Reed canarygrass (*Phalaris aurundinaceae* L.), 32, 39, 41–43, 129, 291
    effect of high-alkaloid content in, 191–92
    effect of seeding depth on percentage emergence, 131
    illustration of, 42
    moisture requirement and soil tolerance for, 122
    percentage of nitrogen recovered as influenced by rate and frequency applied, 233
    seed characteristics for, 135, 137
    structure of eight alkaloids found in, 191
    yield in monoculture and in mixture with orchardgrass, 123
Regal (ladino clover cultivar), 93
Reptans (red fescue cultivar), 36
Respiration:
    in grasses, 145, 147, 148, 149, 150, 155, 159
    following harvesting, 197, 198, 199, 211, 226
    in legumes, 164, 166
Revenue (slender wheatgrass cultivar), 61
Rhizobium, 12, 123, 237, 240
    explanation of, 135

Rhizomes, 5, 7
    definition of, 310
    illustration of, 8
Rhodesgrass (*Chloris gayana* Kunth), 77
    illustration of, 76
    seed characteristics for, 135, 137
Rice, 114
Rongai lablab, 100
Root crops, 119
Rotational grazing, 289, 292–301
    additional costs of fencing and water supply compared to costs of continuous grazing, 299, 300
    changes in the botanical composition, 296–98
    compared to zero grazing, 301
    essential features of, 299, 300
    example of, 293–96
    most important feature of, 293
    principles of, 292–93
    seasonal mean values for herbage constituents and digestibility under, 298
    variations in, 299–300
Rough fescue (*Festuca scabrella* T.), 32
Roughstalked bluegrass (*see* Meadow grass)
Row-crop residues, 114–15
Russian brome (*see* Smooth bromegrass)
Russian wildrye (*Elymus juncea* L.), 45, 48, 97, 148–49, 290
    effect of seeding depth on percentage emergence, 131
    fall utilization for animals given free choice, 60
    germination when harvested by two methods, 128, 129
    illustration of, 47
    moisture requirement and soil tolerance for, 122
    percentage of emergence from four depths of seeding in greenhouse, 48
    in preference studies, 278, 279
    seed characteristics for, 135, 137
    seed production for, 262, 274
Rye, 108, 109, 112, 117
    advantages over oats, 117

Ryegrasses (genus *Lolium*), 43–45, 46, 66, 100, 101, 106, 108, 109, 112, 118
    animal weight gains from bermudagrass pastures overseeded with, 112
    diseases of, 259

S

Sainfoin (*Onobrychis viciaefolia*), 13, 95, 97, 238
    illustration of, 94
    no cases of bloat, 97, 188
    in preference studies, 278, 279
    production and consumption of, 95
    seed characteristics for, 135, 137
    seed production for, 274
    weedy species associated with, 246
Saint Augustinegrass (*Stenotaphrum secundatum* Walt.), 77
Saltgrass, 263
Salvo (timothy cultivar), 52
Saponins, 187
Saratoga (smooth bromegrass cultivar), 32
Scale, definition of, 310
Scarification, 134
    definition of, 128
Seaside bentgrass, 159
Secondary dormancy, definition of, 128
Secondary metabolites:
    classification of, 186
    explanation of, 185
Seedbed preparation, 121
Seed chalcids, 254
Seed-coat modification, 128
Seed germination (*see* Germination)
Seedling vigor:
    characteristics of plants having, 129–30
    definition of, 128
Seed maturity, importance of, 128
Seed multiplication, 273–74
Seed production (*see* Forage seed production)
Seed quality, two aspects of, 121–22, 125–26
Seed trade organizations (*see* Forage seed trade organizations, techniques of)

Seed treatment, 134–39
   inoculation, 135–39
   seed pelleting and coating, 139
Seed types, 273–74
Seminal roots:
   explanation of, 5
   illustration of, 3, 11
Sepal, definition of, 310
Sericea lespedeza (*Lespedeza cuneata* L.), 100, 102, 291
   illustration of, 104
Sessile, definition of, 310
Sheath, definition of, 310
Sheep fescue (*Festuca ovina* L.), 32
Silage, definition of, 310
Silage-making, 196, 215–28
   chemical additives in, 222–23
   dry matter losses in, 226
   harvest equipment used in, 223–25
   problems of, 217–19, 226–27
   process of bacterial fermentation (*see* Bacterial fermentation, process of)
   stages of, 215–16
   types of silos used (*see* Silos, types of)
Silage-making, methods of:
   direct-cut, 219–20
   low-moisture, 219, 221–22
   wilted, 219, 221
Silos, types of, 225–26
   relationship between silo type and moixture content of herbage, 225
Slenderstem digitgrass (*Digitaria pentzii* Stent.), 74
Slender wheatgrass (*Agropyron trachycaulum*), 52, 61, 263
   effect of seeding depth on percentage emergence, 131
   illustration of, 62
   moisture requirement and soil tolerance for, 122
   seed characteristics for, 135, 137
Small-grain crops, 109, 117
   addition of peas to, 114
   crop residues, 114–15
   establishment of, 110
   for grazing, 111–12
   for hay, 111
   for high-quality winter annual pastures, 112
Smooth bromegrass (*Bromus inermis* L.), 29–32, 39, 43, 45, 52, 57, 88, 124, 125, 177, 178, 230, 238, 291
   breeding program for, 270–72
   diseases of, 259, 260
   dry matter production from, 57
   fall utilization for animals given free choice, 60
   fluctuation in carbohydrate reserves, 146, 147–48, 149, 153
   illustration of, 30
   panicle of, 29, 31
   percentage of leaves at various developmental stages for, 180
   percentage of nitrogen recovery as influenced by rate and frequency applied, 233
   physiological stages of, 264
   in preference studies, 278, 279
   proportion of crude fiber when harvested at various stages in growing season, 180
   proportion of crude protein when harvested at various stages in growing season, 181
   response to nitrogen applications, 232
   seed characteristics of, 134, 136
   seed production for, 262, 274
   under rotational grazing, 296–98
   (*see also* Bromegrass)
Smooth--stalked meadow grass (*see* Kentucky bluegrass)
Sodar (streambank wheatgrass cultivar), 61
Sod-bound condition, 232
Sorghum (genus), 74, 82, 108, 109, 114, 118, 125, 159, 186, 248, 299
   danger to livestock from hydrocyanic acid in young, 118
   disease of, 258–59
   grain (*see* Grain sorghum)
   quality characteristics of, 66–67
Sorghum arundinaceum, 118

Sorghum-sudangrass hybrids, 109, 112, 113, 117, 118
    effect of overseeding in a tall fescue pasture, 113
    prussic acid content, 118
Sorgo (sorghum bicolor), 117, 118
Soybeans, 108, 109, 116, 117, 119, 125, 139, 142
    Rhizobium inoculation group, 135–36
    yield of dry matter from soybeans and barley grown in same and alternate rows, 115
Spike, definition of, 311
Spikelets, 7, 9
    definition of, 311
    illustration of, 10
Spring-seeded cereals, 139
Stamen, definition of, 311
Staminal sheath, 12, 14, 15
Starch, 145, 148
Stargrass (*see* Bermudagrass)
Sterling (orchardgrass cultivar), 41
Stigma, definition of, 311
Stink bugs, 254
Stipule, definition of, 311
Stocking rates, determining, 288
Stockpiling, 153, 301, 302
Stolons, 5, 7
    definition of, 311
    illustration of, 8
Streambank wheatgrass (*Agropyron riparium*), 52, 61
    illustration of, 63
    seed characteristics for, 135, 137
Style, definition of, 311
Stylosanthes, 238
Subclover (*see* Subterranean clover)
Subterranean clover (*Trifolium subterraneum*), 102, 106, 109, 112, 119, 190
    estrogenic substance in, 106
    illustration of, 105
Succulent fodders, 119–20
    leaf and stem crops, 119, 120
    root crops, 119
Sudangrass (*Sorghum sudanense*), 66, 108, 109, 112, 117–18, 125, 159, 186, 248, 263, 299
    diseases of, 258–59
    dry-matter yield when treated with dry and liquid manure, 239
    importance of conditioning, 198
    seed characteristics for, 135, 137
Sudax (sudangrass cultivar), 118
Sugar beet, 109, 114
    dry matter yield of fodder, 119
Sugarcane (*Saccharum officinarum*), 109, 114, 117
Sulfur dioxide, 210
Sulfur fertilizers:
    applied to grasses, 234
    applied to legumes, 236
Summer annual pastures, establishment of, 112–13
Summer legumes, 70
Summit (crested wheatgrass cultivar), 57
Sunflower, uses of, 109
Swath, 202
    definition of, 196
Swedes, 119
    dry matter yield of, 119
Sweetclover (*Melilotus* spp.), 13, 81, 88, 97, 99, 101, 109, 119, 133, 136, 238, 262
    antiquality, 99
    as bloat causative, 99, 188
    cause of bleeding disease in cattle, 99, 189
    effect of seeding depth on percentage emergence, 131
    fungus disease of, 257
    importance of conditioning, 198
    moisture requirement and soil tolerance for, 122
    seed characteristics for, 135, 137
    virus disease of, 257
Sweet peas, 15, 265
Sweet sudan grass (sudangrass cultivar), 118
Switchgrass, 66

T

TAC (*see* Total available carbohydrates)
Talladega (crimson clover cultivar), 101

Tall fescue (*Festuca arundinaceae* Schreb.), 32–34, 65, 66, 112, 113, 140, 141, 291
  average annual performance of steers on, 32, 34
  diseases of, 258
  effect of overseeding a sorghum-sudangrass hybrid in pasture of, 113
  germination of, 127
  illustration of, 33
  moisture requirement and soil tolerance for, 122
  proximate feeding stuff analysis results for, 41
  quality characteristics of, 66
  seed characteristics for, 134, 136
  for stockpiling, 301, 302
  toxicity causing fescue foot, 192–93
Tall wheatgrass (*Agropyron elongatum*), 52, 61, 64–65, 159, 194
  effect of seeding depth on percentage emergence, 131
  illustration of, 64
  moisture requirement and soil tolerance for, 122
  percentage of emergence from four depths of seeding in greenhouse, 48
  recurved spikelets of, 61, 65
  seed characteristics for, 135, 137
Tannins, 186, 188–89, 190
Tap root, definition of, 311
Tare (*see* Common vetch)
Tebuthiuron, 142
Tendril, definition of, 311
Testa, definition of, 311
Tetra (alsike clover cultivar), 86
Thiourea, 127
Tifton 44 (bermudagrass cultivar), 72
Tillers, 4, 5, 6
  definition of, 4
  defoliation of, 154–57
  forming stolons and rhizomes, 5, 7, 8
  illustration of, 3
  types produced at base of culm, 5, 7, 8
Timothy (*Phleum pratense* L.), 27, 36, 50–52, 66, 88, 125, 148, 149, 262, 264
  diseases of, 259, 260
  effect of seeding depth on percentage emergence, 131
  germination of, 127
  illustration of, 54
  leaf-stem ratio and proximate feeding stuff analysis for different growth stages, 179
  moisture requirement and soil tolerance for, 122
  panicle of, 52, 53
  percentage of nutrient losses when conserved as hay and silage, 228
  protein production per unit acre of land, 52
  proximate feedstuff analysis of, 174
  quality characteristics of, 66
  seed characteristics for, 135, 137
  weedy species associated with, 246
TNC (*see* Total nonstructural carbohydrates)
Toadflax, 245
Total available carbohydrates, definition of, 146
Total nonstructural carbohydrates, 147
Tripping, definition of, 267
Triticale, uses of, 109
True clovers (genus *Trifolium*), 84–91, 93, 95, 136
  as bloat causative, 99
  Rhizobium inoculation group, 135–36
Turnips, 119
  dry matter yield of, 119

U

Umbel, definition of, 311

V

Vantage (reed canarygrass cultivar), 43
Velvet beans, 108, 109, 119, 291
Velvet bent (*Agrostis canina*), 25
  seed characteristics of, 134, 136
Vetches (*Vicia* spp.), 13, 70, 99, 108, 109, 119, 291
  fungus disease of, 258
  Rhizobium inoculation group, 135–36

Viking (birdsfoot trefoil cultivar), 84
Vitamins, 172
Virus diseases, 257
Vulpia (genus), 32

W

Warm-season perennial grasses, 65–77, 140
    average crude protein content and yield harvested at three-month intervals, 67
    average daily gains and beef production from, 74
    compared to cool-season perennial grasses, 65–66
    diseases of, 258–59
    flowering of, 264
    nitrogen fertilization of, 231, 232
    quality characteristics of, 66–67
    response to potassium fertilization, 233
    tolerance of soil acidity, 238
Weeds:
    avoiding infestations, 252 (*see also* Companion crops)
    characteristics of, 245–46
    control methods, 248–52
    definition of, 245
    losses due to, 244
    poisonous, 247–48
    species frequently associated with some common forage species, 246
Weeping lovegrass, 263
    seed characteristics for, 134, 136
Wheat, 108, 109, 112, 114, 117, 159
    grain yields grown at three ryegrass densities, 139, 140
    yield of dry matter from wheat and soybeans grown in same and alternate rows, 115
Wheatgrasses (genus *Agropyron*), 45, 52–53, 55–65
    diseases of, 259
White clover (*Trifolium repens* L.), 12, 19, 27, 36, 70, 72, 86, 91, 93, 99, 100, 101, 125, 162, 166, 186, 238, 265, 291
    as bloat causative, 188
    fungus diseases of, 257
    for heavy grazing, 289
    illustration of, 13, 90, 91
    in preference studies, 278, 279, 280
    seed characteristics of, 134, 136
    types of, 91
    virus disease of, 257
    weedy species associated with, 246
White-fringed beetles (*Graphognathus* spp.), 254
White lupine, 13
White man's foot grass (*see* Kentucky bluegrass; White clover)
White sweetclover (*Melilotus alba* Deor.), 97, 99
Wildrye (genus *Elymus*), 45, 47–50
    diseases of, 259
Windrow, definition of, 196, 311
Windrowing, process of, 196–97, 200, 201, 203, 223
Winter annual pastures, establishment of, 112
Winter legumes, 70

Y

Yamhill (wheat cultivar), grain yields grown at three ryegrass densities, 140
Yellow sweetclover (*Melilotus officinalis* L.), 97, 99
    illustration of, 96
Yukon (yellow sweetclover cultivar), 99

Z

Zerna (subgenus), 29
Zero grazing, 300
    compared to rotational, 301
Zim (timothy cultivar), 52
Zygomorphic, definition of, 311